An Environmental Life Cycle Approach to Design

John Cays

An Environmental Life Cycle Approach to Design

LCA for Designers and the Design Market

 Springer

John Cays
New Jersey Institute of Technology
College of Architecture and Design
Newark, NJ, USA

ISBN 978-3-030-63804-7   ISBN 978-3-030-63802-3   (eBook)
https://doi.org/10.1007/978-3-030-63802-3

This Springer imprint is published by the registered company Springer Nature Switzerland AG
The registered company address is: Gewerbestrasse 11, 6330 Cham, Switzerland

To my family—my parents, John Bennett and Victoria Torres Cays, who taught me and my brother Stephen, and sisters Elizabeth and Kathrine, that there is always enough; my wife Angela who patiently walks with me on the path of living each day simply and sanely; and my daughter Sara who is my absolute favorite person on the planet. Thank you all for giving this journey meaning and joy.

Preface

The designer's primary job is to present compelling synthetic solutions that meet people's needs. Good design, at every scale, wins both hearts and minds. It can stir the emotions while satisfying multiple utilitarian requirements. At this point in human evolution, however, it is necessary—and possible—to do more.

Every 4 days, a million babies are born to families around the world. As more and more people strive for better lives, how do we design a world that not only significantly reduces our already outsized global footprint but also changes negative environmental impacts into positive ones? This is arguably the largest design challenge in the history of the world.

Design professionals entrusted with protecting public health, safety, and welfare, as their primary public charge, can now shape humankinds' response to this challenge. To practice ethically, we must now evaluate each new product's potential to physically impact the earth's ecosystems *throughout its entire life cycle prior* to bringing it into the world. How our species survives and thrives throughout the next 50 years and beyond depends on the individual and collective choices we make about how and what we design, make, and consume.

Science-based design approaches can positively influence those decisions. Life cycle assessment methods and tools can provide a clearer picture of how to optimize individual designs that not only do less harm but also promote healing in our ecosystem. This book clarifies the terms and root causes of the environmental dilemma we face today—the result of market responses to *both* rational and irrational human demands—and presents life cycle assessment (LCA) within this context as an illuminating, data-driven methodology to help designers and their clients make better, evidence-based decisions.

In order to affect positive environmental change, it is not enough to provide facts. Data quality increases our abilities to precisely describe and present realistic solutions to problems that humankind presently faces. Change is possible, but only when design operates with an awareness of the larger context, that is, the constant psychological, political, and economic forces at work. Arguments fueled by fear, mistrust, greed, and inertia, made by the very people for whom we are designing, can undercut facts and best intentions to sustainably meet the world's design challenges.

Effective and iconic designs typically emerge as elegant manifestations that move people. Today, they must also function to heal the planet. This will require all the ingenuity our species can muster, and it cannot happen without establishing an underlying system, like LCA, to track our progress toward the goal of a truly sustainable marketplace.

That said, design is primarily a visual discipline. Many practitioners, having been educated in schools of art and design, may not be naturally inclined to embrace such technically driven design methods, preferring instead to rely on more heuristic or intuitive approaches. This book will introduce designers to a complementary set of science-based perspectives and techniques that address a global market that increasingly demands proof that things perform ecologically as advertised. Just as good design wins minds by solving people's everyday problems and wins hearts through solutions that move beyond the mere utilitarian to inspire delight, LCA data-supported design solutions win minds through proof and hearts through trust.

While this book will provide some detailed information about LCA, and refers to the latest research in the field, it is not a technical compendium. Rather, it is a bridge for those in the design fields who are uninitiated to but interested in how LCA is helping to solve the grand environmental challenges of today and tomorrow through data-driven, science-based techniques. Since our existential challenges are not merely technical but are also rooted in misperception, misinformation, fear, and confusion, some chapters in the book contextualize the problems for which LCA and related analytical and visualization techniques can provide solutions. Other chapters work to cast LCA as a normal part of the designer's workflow.

Designers and their clients are primarily focused on appearance and performance; environmental impact issues are minor considerations relative to cost, utility, and aesthetics. Billions of consumers playing by traditional rules of the marketplace will only accelerate ecological destruction. Environmentally responsive designs may never be realized if they are cost prohibitive. Complex and expensive regulatory or product certification systems create excuses for not bringing the best solutions to market. Smoothing the path for designers to easily "bake in" invisible ecological benefits to everything they do, at little or no extra cost, requires awareness, understanding, and evolution in twenty-first-century tools and work habits. This book is a guide and an exhortation to work smarter, provide added value to clients, and to make these changes now.

Newark, NJ, USA   John Cays

Acknowledgments

I would like to express my sincere gratitude to my colleagues at the New Jersey Institute of Technology and the Hillier College for creating and supporting an atmosphere of free inquiry. To Dean Urs Gauchat who first encouraged me to look into the topic of life cycle assessment and how it might benefit the world in the hands of designers. He insisted that I set aside dedicated time each day to learn and write about this topic to help make the planet better for my daughter, her friends, and their children. To Dean Tony Schuman who continues to serve as an example of an activist scholar. To Dean Branko Kolarevic who enthusiastically made sure I completed the mission. To Provost Fadi Deek, who always gave gentle encouragement to persist in writing a first book.

Special thanks go out to David Brothers, Martina Decker, Christine Liaukus, Darius Sollohub, and Andrzej Zarzycki for their sincere interest in the subject of LCA in design and their perennial invitations for me to introduce it to their design students. Also to Glenn Goldman who encouraged me to think broadly, be curious, and to see the vital connections between design and technology. To Maya Gervits who could always anticipate my questions and needs on which library resources would best lead me to fertile areas of exploration, and also for the generous support she provided with library resources on extended loan.

To Ed Mazria who, well over a decade ago, set the aspirational goal for every student in every university design studio to have a constantly updating digital readout on the ecological impact status of their designs, and to the tireless work of the LCA researchers and practitioners without whose work over the past five decades this would be impossible to achieve.

To my sangha and my teachers, who continue to teach me to be mindful of and advocate for all the vitally important things that cannot be seen. To Cynthia Imperatore who always reminded me that people need more than technical facts to care about something. You are so missed. To Laura Peterson and Lou Kilgore for keeping me going every Sunday.

To the American Center for Life Cycle Assessment community for inviting a wider set of contributors to add their voices and ideas to broaden LCA adoption. Your encouragement, leadership, and messaging makes a tremendous difference.

To all the members of the Springer editorial team who have been most patient and supportive as life and work prolonged delivery of a final manuscript.

To all the design students in my classes over the years who asked honest and earnest questions to learn and practice how to back up their sustainable design claims. And to the skeptical ones who needed to dig deeper to know how we know what we know about our individual and collective impact on this living world. I especially want to thank Erin Heidelberger who followed farther than any Hillier design student and tenaciously persisted in an entire year of rich and challenging dialogue. She appeared at the perfect time in her own soft-spoken but determined way to make meaningful contributions to this book through revisiting a major design studio project, her illustrations, and detailed research on how designers and design students can actually *use* LCA. She is well on her way to help design make a *measurable* difference.

Contents

About the Author

John Cays is Associate Dean for Academic Affairs and Interim Director of the School of Art + Design in the New Jersey Institute of Technology's J. Robert and Barbara A. Hillier College of Architecture and Design. He is a licensed professional architect. He holds a Bachelor of Science in Architecture from the University of the Arts and Master of Architecture from Princeton University.

Prior to cofounding GRADE Architects, an architecture and interior design firm in NYC in 2001, he was a project manager at Robert A.M. Stern Architects. Since 2005, he has been responsible for overseeing the development and use of "Kepler," NJIT's transparent digital repository, and the qualitative data-driven design curriculum management and assessment system. In 2008, Kepler served as the engine behind the nation's first fully digital NAAB accreditation visit. He served as North East Regional Director for the Association of Collegiate Schools of Architecture from 2014 to 2017 and as Director on the National Architectural Accrediting Board from 2017 to 2020. He is an active member of the ACLCA Education Committee.

His research focuses on visualization, translation, and advocacy to increase the adoption and use of quantitative life cycle assessment methods and tools in the design fields.

Chapter 1
Do Nothing: The Danger of Believing in a World Without Limits

Abstract Human desire and ingenuity have shaped the modern world. The natural human need to consume and fears of scarcity drive desire for more and more creature comforts. We must and can improve the way we consume.

The *American Way of Life* is often interpreted as an exhortation to freely pursue individual happiness without regard for physical limits or finite resources. Engendered at the end of the eighteenth century in a new nation, this worldview is rooted in the pragmatic and powerful impulse to satisfy individual well-being at a time when nature's abundance seemed limitless. That idea continues to beckon us to live large. It is still the context in which we consume nearly two-and-half centuries later.

The penchant for planning, making, and consuming is normal; it is what humans do. Yet as technological innovation has removed nearly all natural limits to our species growth, it has also produced myriad, unintended, negative environmental consequences. Reversing these effects requires us to coordinate "best practices" to guide our individual and collective decisions.

New and constantly evolving tools, data, and methodologies reveal our impact on the planet as we design each new product and service. They allow us to quantify, evaluate, and improve the environmental impacts. It promotes system-wide approaches to innovate truly sustainable ways to satisfy growing global demand.

1.1 Living Large: The American Dream

Some of us live large. Not content with mere survival, we take in and give back significantly more than we need to maintain internal physiological balance—homeostasis—in our environment. Every day, more people around the world embrace an ethos of overconsumption. Fueled by imagined, personal, future gains, it is a hedge against suffering that expresses itself as acquisition, a desire for more and more stuff. The starkest expression of this ethos is, perhaps, the aspirational American dream. No longer confined to the culture that coined the term, it is magnified by at least three orders of magnitude as continues to spread across the globe.

While more than half of the earth's seven billion people live on less than $10 per day (Rosling et al. 2020), the number of people able to participate in acquisitional pursuits, representing individuals in every country and culture, is still measured in the billions. Meanwhile, the other half live in relative poverty. Forced to live a frugal existence, they demonstrate everyday what the human body needs to survive, the basics—food, water, shelter, and the most basic transportation means. But all people dream. In a connected world, comparisons are ever present and beckon those without to join in the pursuit of more and better.

As we strive to do better on and for our finite planet, it is impossible to take fundamental human needs *and human desire* out of the equation. As designers, charged with manifesting those desires in the marketplace, we can do better for everyone. But without reimagining and redefining what living well means and how more people can reach that standard, our current models will be increasingly unable to support the "self-evident truths" espoused a quarter millennium ago by the framers of the US Constitution. Without a detailed consideration of specific means and methods to sustainably promote them, Life, Liberty, and the Pursuit of Happiness will increasingly diverge as achievable ends for almost everyone.

1.2   The American Way of Life

"Way of life" is a phrase that is, at once, fluid and solid. It implicitly acknowledges processes that respond to conditions along a variable path and, at the same time, forms a fixed, monolithic idea. Our way of life emerges each moment and evolves through the choices we make. It is an organic expression of what is possible within the complex set of perceptions, circumstances, and forces that simultaneously allow and limit its progress. The way emerges from its conditions just as a river flows through terrain, guided by forces of gravity and friction that work as part of much larger cycles.

In the United States, politicians from both dominant parties invoke "the American Way of Life" as a solid and singular thing that we must, at all costs, protect and preserve. Coined by James Truslow Adams, the term captures such concepts as "bigger and better" and "a better, richer, and happier life for all citizens of every rank" (Adams 2017) and has provided a powerful metaphor throughout the twentieth and into the twenty-first centuries. Barak Obama launched his 2008 presidential bid by sharing his "thoughts on reclaiming the American dream" and extending it to people who felt they had lost or never had the ability to realize it (Obama 2006). Eight years later, Donald Trump built an ultimately successful campaign on promising his electoral base all the material trappings of what it represents. The candidate, who is out of touch with how far people feel they are from attaining all its promises, risks losing an election. It has proven to be a powerful ideal that is firmly set in the minds of most people, not only in the United States but also around the world. It implies both expectations and behaviors of free individuals earning a livelihood and pursuing happiness.

1.2.1   The Pursuit of Happiness

Written into the first sentence of the Preamble to the US Declaration of Independence, "the Pursuit of Happiness" is an aspirational and beguiling principle that has propelled the American Way of Life ever since it was penned. It promises every citizen the possibility of attaining *security, comfort,* and *prosperity*. What has followed, since the founding fathers codified the powerful open-ended notion in the summer of 1776, is an unrelenting pursuit that continues to manifest as consequences and byproducts.

"The Pursuit of Happiness" as a key principle is most compelling perhaps because it does not consider any external effects or limits. It says nothing of means or costs to be considered in the pursuit. It was written at a time when the world seemed limitless and foresaw only benefits to both the individual and the growing community through the development of new enterprises. A steady increase in negative ecological impacts, however, closely correlates with the burgeoning wealth that was created from the end of the eighteenth through the start of the twenty-first century.

1.3   No Going Back: Impossible to Do Nothing

Many blame modern political, social, and economic structures for all woes and suffering on the planet. Some advocate for a complete "reset" through various means, implemented either gradually or suddenly. It is, however, impossible to stop or move backward to "a simpler time." Even if every one of the nearly eight billion people alive today were to agree to simply stop, to abandon the chase altogether in order to reduce our collective impact on the planet to zero, we must realize that it is impossible to actually *do nothing*.

To be requires the human body to ceaselessly do something. Even before we are born, we are each, already, doing something. From soon after conception until we take our last breath, we take oxygen and nutrients into our bodies and expel waste. Sitting perfectly still, even sleeping, the body must consume energy and resources to live. All human beings alive on the planet today *take in and give back* each second of each day we are alive and beyond. Only death can stop our ceaseless consuming and release the stored biochemical energy in our bodies to dissipate back into the larger ecosphere over time. An incalculable number of other organisms, large and small, inhabiting the land, sea, and sky, also take in and give back. It is what defines being alive and part of the earth's *ecosphere*.

1.3.1 Humans in the Ecosphere

The earth's ecosphere, of which we are a small but disproportionately active part, is a self-contained system powered by the sun. This interconnected network and its support system reaches through the rock and soil under our feet, permeates the deepest ocean waters, and extends high into the thinnest air. This system maintains a dynamic equilibrium. It self-regulates through short weather cycles as well as longer climate patterns. As these natural cycles and patterns shift and change over time, life becomes easier for some and harder or impossible for others. We are one of millions of species that have evolved, adapted, and survived as the earth has cooled or warmed throughout the ages.

Several distinguishing behaviors make us, perhaps, the most adaptable species to a changing climate of all life forms alive today. We *think*. Abstract ideas, concepts, concerns, and plans occupy nearly every waking hour. We *express* ourselves. We communicate our thoughts, feelings, and perceptions about our world and our place in it through signs and symbols. The largest and most lasting physical impact, though, comes from how we *act* and what we *make*.

Our collective actions physically transform the ecosphere. The things we make persist through time to form a complex support infrastructure called the technosphere. As an artificial subset of the ecosphere, it disrupts and delays the impact of natural forces to break apart and recycle things back into constituent elements and compounds in forms typically found in nature.

1.3.2 Creating the Technosphere (30,000 BCE)

Over the last trillion seconds, we humans have been recording and sharing our experiences; our ancestors drew the earliest records of human activities in caves a little over 30,000 years ago. Since then, we have formally manifested our ideas through an increasing array of physical and, very recently, virtual media. We started in earnest *only* around a trillion seconds ago, setting ourselves apart from the rest of nature by not only manipulating nature to satisfy our individual and collective desires but also by recording the details of what and how we lived and did things using the most durable media available to paint on the walls of caves (see Fig. 1.1). Thus, our growing technical abilities have allowed us to become much more than a simple collection of individuals progressing through our own *purely* biological life cycles. Our species' ambitions and appetites have grown over the millennia along with our technical means satisfy them. We continue to invent new ways to provide for our growing numbers. As our ability to feed, clothe, house, medically treat, communicate with, educate, and entertain ourselves increases, so does our collective rate of taking in and giving back—with significant negative implications for us and for the interconnected web of living beings.

Fig. 1.1 *Fragment of* prehistoric cave painting using earth pigments in the "Hall of the Bulls" at Lascaux (28,000–17,000 BCE). The cave is located in the Aquitaine region of France, in the Dordogne department in the commune of Montignac. (Image courtesy of Peter 80 (2005), CC Attribution-Share Alike 3.0 Unported license)

Fig. 1.2 The Black Marble. Lights burn steadily every night as glowing embers on a continental scale. (Image Credit: NASA Earth Observatory image by Joshua Stevens, using Suomi NPP VIIRS data from Miguel Román, NASA's Goddard Space Flight Center (2016))

As humans, we continuously develop new technologies that allow us to ever more effectively magnify our ideas and turn them into objects, large and small, in the physical world. Mastery of science, economics, and design and a steady supply of abundant energy and raw materials allow us to turn pockets of night into day (Fig. 1.2) and have allowed some to realize their most complex plans in order to make and distribute nearly anything wherever and whenever they want at massive scales and in annual quantities counted in the billions and even trillions.

Every invention comes from the rational impulse to address a real human need. Each thing we synthesize from natural elements, then use and dispose of, comprises and then "lives" in the *technosphere,* the sum total result of all human activity on the planet. We constantly remove and transform elements from the ecosphere. We harness the embodied energy of animals, plants, and minerals for our own use every day. The instant we tap any biological, geological, hydrological, or atmospheric element as a resource to make something, it begins its own *artificial life cycle that* ends when it either gets repurposed or eventually completely "metabolized" by natural forces back into the ecosphere.

1.3.3   The Holocene (11,700 Years)

Over approximately the last 12,000 years, since the end of the last Ice Age that roughly defines the Holocene Period, icy glaciers retreated and human civilization flourished. As human populations grew, migrated, and permanently occupied an increasingly ice-free world, so did human ideas and technology. Individual inventions multiplied and coalesced into two intertwined parts of our technosphere. One is composed of supportive and useful objects, the celebrated inventions that make life safer and more comfortable; the other is the sum total of what we have tried, unsuccessfully, to throw away, our waste.

We are all now dependent on massive and inherently complex systems built to support and improve human life. As appealing as it may be to some, there is no quick transition to simply stop and go back to "a simpler world" devoid of modern technology without increasing physical suffering and shortening the lives of most people living today. Watching what happens when the power goes off for a week or even a few days in our metropolitan areas gives insight to the difficulties that arise. Without the energy and communication infrastructure on which civilization has come to rely, it can mean the difference between life and death for some (Anderson and Bell 2012) Rather than attempting to radically eschew complexity and its inherent challenges and contradictions, the key to our surviving and prospering on this planet is our continuous and successful technological adaptation to changing conditions.

This is not to say that all technology is inherently good (or bad) or that we can suspend thoughtful judgment on the long-term implications of our actions. The technosphere is now inextricably woven into the structure of the ecosphere. It is now an artificial "second nature" to those of us who depend on its systems to support our habits and choices concerning where and how we live in a contingent and uncertain world.

1.3.4   Human Civilization (2500 Years)

The codependent and ephemeral nature of human existence has been a central tenet in both Eastern and Western philosophical traditions from Classical Antiquity. From Hinduism's concept of Samsara (Rodrigues 2006) to the teachings of the Buddha on impermanence (Hanh 1999) and Heraclitus's doctrine of universal flux (Heraclitus 2020 Patrick trans.) for the last quarter of the Holocene interglacial period (2500 years), as human civilization has developed and human thought has evolved, so has the understanding that *change* is the one constant on which we can depend. Civilization has simultaneously built increasingly stronger hedges in an attempt to protect ourselves from life's uncertainties.

We have developed practices, habits, and complex systems, through which we relate to the unpredictable and hostile natural world. Large-scale technological advances in areas including agriculture, energy production, medicine, transportation, and building systems are effective guards against discomfort, insecurity, and scarcity; however, successfully mitigating the vicissitudes of an unmediated existence in the wilderness comes at a price. As these artificial systems have acted to stabilize the conditions in which humankind can survive and grow, they have also accelerated the pace and increased the degree of change in natural systems during this same period.

1.3.4.1   Increases in Population Magnify Impact

As our numbers multiplied at least 50 times over the same 2500 years, from approximately 125 million globally (roughly the size of Japan's current population) to the current number approaching 8 billion, our geographic reach spread over every continent and climate. The ingenious solutions that have allowed us to survive and thrive in places where we are not biologically adapted to live come at a cost. Living in these places is made possible, almost entirely, by extracting natural resources and setting them on fire to generate energy (Fig. 1.3).

This continuous extraction and burning has taken place for hundreds of millennia since we were living in caves. It started the moment our ancestors first harnessed fire. Our latest evolutionary stages were influenced by the fact that they could gather and do things beside a fire where and whenever they chose to build one. For a half million years or more, according to infrared dating techniques, even before our species evolved into modern *Homo sapiens*, our forbearers collectively demanded more of the physical comfort and security that fire provides (Berna et al. 2012). This also allowed them to move into areas that were too cold to inhabit without a controlled and dependable heat source.

Fig. 1.3 A coal power plant in Datteln, Germany. Per capita energy consumption rose by 33 percent between 1990 and 2005. Between 2005 and 2030, per capita energy consumption is expected to increase another 18 percent, meaning that global consumption will rise by 50 percent. The majority (86 percent) of the world's energy needs are met by fossil fuels. (Photo Credit: Arnold Paul | Wikimedia Commons)

1.3.4.2   Urbanization

More than half of the world's population live in cities today. As the world continues to urbanize, in order to support billions more people, we will do nearly anything to continue keep the fires burning for our families and ourselves. Viable human settlements have always needed fire in order to exist. But the type, quantity, and byproducts of the "fire" we now use have far-reaching environmental implications.

Without the continuous burning needed to house and feed people; to heat and cool our homes; and to make, transport, and store our daily necessities, every major city in the United States today would be either uncomfortable or completely uninhabitable for at least part of the year.[1] As the burning continues and increases to power nearly every growing hamlet, village, town, and city around the globe, we are witnessing global-scale changes affecting the fundamental conditions on which most life depends.

---

[1] The only exceptions of which I am aware may arguably be San Diego, Los Angeles, and Long Branch, CA. These Southern California cities current temperate climate could support a small fraction of the US population without relying on the fruits of technology to maintain year-round thermal comfort. However, abundant fresh drinking water, a steady food supply, advanced medicine, and all other technology-dependent basic systems would still require energy to function.

1.4 Relative Comfort and Quality of Life

Once basic needs are provided for, the bar that marks minimally acceptable creature comforts quickly rises. Each additional requirement to elevate an individual or family above a level of subsistence living with minimal heat, food, and shelter adds to the number of goods and services needed to satisfy them. Each comes with an additional energy cost and attendant harmful environmental emissions. The relative increase to the store of material possessions and replacement when they are no longer fit for purpose is largely what comprises our Pursuit of Happiness today.

Underpinning and constantly operating below the surface of this culture of pursuit is a subtle psychological process that keeps most of us comparing our relative positions in society, in either obvious or subtle ways (Dubois and Ordabayeva 2015). It is at play, whether we examine it or not.

Karl Marx uses almost exactly the same example of two adjacent houses to illustrate how ego and striving for recognition fuel the desire to acquire and the impulse to consume conspicuously:

> A house may be large or small; as long as the neighboring houses are likewise small, it satisfies all social requirements for a residence. But let there arise next to the little house a palace, and the little house shrinks into a hut. The house now makes it clear that its inmate has no social position at all to maintain, or but a very insignificant one; and however high it may shoot up in the course of civilization, if the neighboring palace rises in equal or even in greater measure, the occupant of the relatively little house will always find himself more uncomfortable, more dissatisfied, more cramped within his four walls. (Marx 2008)

Comparisons

In the summer of 2014, my 4-year-old daughter, Sara, walked up to me in our yard and reported, "Becky says we live in the tinyeeeest house." My little girl was looking at me with one eye through the sliver of space between her forefinger and thumb as she repeated what her next-door friend had said. She didn't seem bothered by the observation, but I felt compelled to respond with something like:

"Most people live in houses that are bigger than some and smaller than others," I said, "and for a long time humans didn't live in houses at all." That brief exchange started me thinking about the decisions that had led my wife and I to this leafy enclave in Northern New Jersey and why, ultimately, we bought this particular small one-and-a-half-story Cape Cod in the town of Wyckoff. Although the public schools were highly rated and the taxes reasonable by New Jersey standards, what we found most captivating was the street, neighborhood, and the sense of community with its relatively densely packed post-World War II housing built for returning GIs and their growing families.

When we first visited on a sunny summer day, with kids running around in neighboring yards and a nature reserve just around the corner, it seemed like a great place to put down roots to raise our newborn. The south sun flooded the living room of the corner lot at the mouth of the cul-de-sac all day long, and in 2009, after the bottom had dropped out of the housing market, the price was right. Although most of these 1950s tract homes had been remodeled, and many enlarged somewhat, the street had retained a cohesive feel. It was a little piece of the middle-class ideal, where, with all the smallish working-class houses on quarter acre lots, we wouldn't be distracted by people trying to "keep up with the Joneses."

From the neighbor girl's vantage point, however, in her moderately larger house with a pool in the yard, I guess it did seem tiny. Becky's comment was an innocent and honest assessment of the relative size of things. It's what we all do as kids. We try to figure out our place in the world and make sense of things by comparing our own conditions to those immediately adjacent to us. As we grow up, we retain the impulse to compare. It colors our perception of what is enough, what we lack, and what we demand in order to fulfill our needs. Comparing can be an unhealthy, human impulse that drives consumption. When left unexamined and unchecked, our natural appetites for more, bigger, and better can easily grow to colossal proportions.

This insight on fundamental human nature is as valid today as it was when it was delivered in the mid-nineteenth century. The illustration applies to any moment where the simple act of comparing fuels an endless race to make and consume things. The human impulse to satisfy personal physical, emotional, and psychological needs and to judge the level of this satisfaction or "happiness" in a social context is universal. It establishes the baselines by which we measure "standard of living." It feeds the perceived need for constant growth, production, and development everywhere in the world. It is played out in each decision we make from what to eat, how to travel, where to live, and how to spend each day.

1.4.1 Standard of Living Disconnected from External Impacts

The notion of a mother and father raising two children in a house with a yard in the suburbs, as the aspirational pinnacle of the American Way of Life, has been effectively challenged since it reached its full expression in the middle of the twentieth century. This is evidenced by the many alternatives that exist today, which have redefined models of domesticity, personal relationships, and family. While such changes in social structures may be more inclusive and psychologically healthier for more people, *they generally have materially the same per person ecological impact as previous structures.* Whether living in a city, the suburbs, or in a rural village, the basic necessities of comfort and security are the same. In order to live we all need food to eat, clean water to drink, and a warm dry place to come home to.

In order to fairly compare our choices of *where* we decide to live, when it comes to environmental footprints and sustainability, we must consider *what else* will support that primary decision or occur as a result of it. For instance, if we live in the suburbs or in rural enclaves, we will need one or multiple cars. Some proponents of a spatial solution to ecological problems argue for density, for living in cities. If you live in city, you don't need car. But we have look more closely. Cities typically provide more convenient access to a different, and often more varied, kind of goods and experiences. Circumstances and proximity of neighbors, and the number of neighbors you have to compare to, tend to drive consumption to differentiate your "station" and require higher-order consumption. People in cities tend to go out more, to partake in culture in different ways than those in rural or suburban areas may not. These pursuits, supported by robust physical and information infrastructure, require additional energy, which adds to the overall net expenditure of energy and also increases levels of damage (within an economy that is not fully measuring or aligned to measure damage).

While standard of living tracks certain material and quantifiable metrics, there is no single objective "standard of living" against which the relative positions of all people are measured. The term refers, rather, to "the amount and quality of material goods and services available to a given population. The standard of living includes basic material factors such as income, gross domestic product (GDP), life expectancy and economic opportunity (Silver et al. 2019). It generally refers to wealth, material goods and necessities of certain classes in certain areas" (Fontinelle 2020).

Most of us in the United States are accustomed to a certain standard of living that we would miss if it suddenly, or even gradually, disappeared. If *forced* to change routines and habits overnight, so that we used only what we physically need to survive, the resulting social and political strain on the collective would be enormous and could end in violent revolt. Turning the haves into the have-nots in order to be able to reduce our collective ecological footprint and live sustainably on the one planet we have generally works only when it is someone else who sacrifices and goes without.

No matter where we live, we each have our individual routines and habits. We each tend to procrastinate, deny, or resist when faced with the need to alter those routines and habits, to give up the material trappings of our individual way of life. When faced with making such potentially disruptive personal decisions, it is not hard to see how protracting the debate on the roots of climate change receives tacit support (Gutstein 2018). Some of us delay and hesitate to act. We wait for more conclusive scientific evidence on the particulars, for a perfect and irrefutable computer model of the earth's climate system, and for statistically certain methods with more and more data to guide our decisions. Until we have perfect knowledge in place, we can conveniently maintain and even expand the status quo that underpins our established *way of life*.

Each personal decision on how to satisfy needs and wants sets in motion thousands of interconnected reactions. From the shoes we each wear to the video we live stream, global markets anticipate and seek to satisfy these demands. The net result from the constant subjective state of having or being *less than* ends in our making more stuff to equip ourselves to better shield ourselves from threats, real and perceived, and constantly improve our living standards. Satisfying growing global demand for goods and services takes ever-increasing energy. The predominately carbon-based petrochemical sources of energy used to produce with environmental impacts that have now, by many objective measures, are reaching or some say have reached a tipping point (Wackernagel et al. 2002) and require careful monitoring going forward.

Basic needs are not particular to Americans. What distinguishes the *American* way to meet these needs that define a baseline standard of living lies in the particular combination of unalienable rights formalized at the top of the US Declaration of Independence. It first recognizes and legitimizes the strong human urge to survive—*Life*—then safeguards individual freedom to choose the means of satisfying that fundamental urge—*Liberty*—and finally promotes the ideal that a better future is achievable through determination, perseverance, and hard work—*the Pursuit of Happiness*. A threat or perceived attack on any of the three components in this firmly established notion elicits strong emotional responses from those who take them to be guaranteed rights.

The American Way of Life has, up to now, been resource intensive. It is designed to be that way. The lifestyle that defines basic living standards in the United States and other high-income societies requires the transformation of excessive amounts of natural resources into consumer goods and experiences. In return, we produce excessive amounts of waste that has to be put someplace. Understanding how this works is the foundation for a steady evolution of a better-designed way of life—one product, one building, one experience at a time.

1.5   Sustainable Development

The roots of the environmental movement have decades-long history in the United States but have always been challenged by a national ethos of overconsumption that has an even longer history. Any response we make as designers to try and improve how and what we make must be mindful of the underlying market forces and political agendas that conspire to maintain the status quo. The notion of "pursuing happiness," written into founding documents of our nation, has been used for generations to support material acquisition, and every generation's ideal of a reasonable living standard has grown to include more and better goods and services.

Today, design works at the crossroads of supplying increasingly insatiable market demands for better and more while effectively balancing conflicting ecological, economic, and socially equitable stakeholder interests. Working this way is imperative to any truly sustainable design agenda. Sustainable development includes everything new, at every scale, that we bring into the world to satisfy real or perceived needs. The generally agreed upon definition was concisely articulated in the Brundtland Report, *Our Common Future* by the World Commission on Environment and Development commissioned by the UN General Assembly in 1987. It defines sustainable development as that which "meets the needs of the present without compromising the ability of future generations to meet their own needs" (Brundtland et al. 1987).

The report identifies "interlocking crises" and points out that "Until recently, the planet was a large world in which human activities and their effects were neatly compartmentalized within nations, within sectors (energy, agriculture, trade), and within broad areas of concern *(environment, economics, social).* These compartments have begun to dissolve... These are not separate crises: an environmental crisis, a development crisis, an energy crisis. They are all one" (Brundtland et al. 1987).

This laid the groundwork for the work published a decade later in 1997 by sustainability expert, John Elkington, who coined the useful term "triple bottom line" to describe a practical approach for corporations to optimize over multiple conflicting criteria. He puts it this way: "Sustainable development involved the simultaneous pursuit of economic prosperity, environmental quality, and social equity. Companies aiming for sustainability need to perform not against a single, financial bottom line but against a triple bottom line" (Elkington 1998).

When designers follow these principles, truly sustainable, triple-bottom-line, considerations can move to the early stages in project and product life cycles. Life cycle assessment (LCA) is one proven method that uses tools and information to evaluate the impacts of a design across multiple environmental impact categories. Emerging datasets and digital interfaces that focus on the financial and even to less available, social impacts can also be incorporated into the same life-cycle-oriented design workflows to provide designers insight into a triple-bottom-line design approach. A full treatment of financial cost and social impact bottom line issues arc beyond the scope of this book, but the environmental life cycle framework can be directly applied to these challenges.

1.6 Measurement

Design responds to market demands, driven by human nature, everywhere in the world. Each culture has its own governing norms that influence what their people ultimately demand to support their own standard of living—but all people wish to live happy, safe, and free. Liberty is defined by a person's ability to freely move beyond boundaries. The illusion of extreme individual liberty as a protected right to say, do, or experience anything without limits comes at a price when multiplied by millions or billions of people with increasing means to consume as an individual "expression" of style or taste.

As more of us continue to engage in this pursuit, irrespective of the means, the consequences of the incessant chase have reached a point that requires close monitoring and careful study. The need to accurately measure and record the environmental impacts of our way of life, of individual decisions to buy and sell, to do or make something, has brought forth new frameworks for assessment.

Later chapters present and discuss the methods and tools that help designers monitor the environmental impact created by the things we are asked to design to meet market demands. These tools and techniques are predicated on setting boundaries according to project goals and scopes. These boundaries have nothing to do with curtailing personal liberty or even recognizing limits of resources. They are, rather, required to define a system under study and, as such, should be seen as part of an objective system to monitor and document the flow of resources. This assessment system can also be integrated as part of a workflow in design professions ethically bound to safeguard the health, safety, and welfare of the public.

The impacts of *how* we live and *what* it takes to support our standard of living can now be quantified. As individual behaviors and societal norms continue to evolve over the next two generations, we will benefit from new technologies that will tally their environmental effects on the fly. Our assessment tools can already semiautomatically keep score and evaluate the risks from subtle to major shifts in how we consume energy and material resources as we pursue happiness in all its forms.

References

Adams JT (2017) The epic of America, Illustrated Edition. Routledge, New York, NY, pp xx

Anderson GB, Bell ML (2012) Lights out: impact of the August 2003 power outage on mortality in New York, NY. Epidemiology 23:189–193. https://doi.org/10.1097/EDE.0b013e318245c61c

Berna F, Goldberg P, Horwitz LK et al (2012) Microstratigraphic evidence of in situ fire in the Acheulean strata of Wonderwerk Cave, Northern Cape province, South Africa. Proc Natl Acad Sci U S A 109:E1215. https://doi.org/10.1073/pnas.1117620109

Brundtland GH et al (1987) Brundtland: our common future (report for the world... – Google scholar). Oxford University Press, Oxford, pp 13–41

Dubois D, Ordabayeva N (2015) Social hierarchy, social status and status consumption. In: The Cambridge handbook of consumer psychology, Cambridge University Press, Cambridge, England, pp 332–367. https://doi.org/10.1017/CBO9781107706552.013

Elkington J (1998) Cannibals with forks: the triple bottom line of 21st century business. New Society Publishers, Gabriola Island, BC/Stony Creek, CT

Fontinelle A (2020) Standard of Living vs. Quality of Life: What's the Difference? In: Investopedia. https://www.investopedia.com/articles/financialtheory/08/standard-of-living-quality-of-life.asp

Gutstein D (2018) The big stall: how big oil and think tanks are blocking action on climate change in Canada. Lorimer, Toronto, p 86

Hanh TN (1999) The heart of the Buddha's teaching: transforming suffering into peace, joy, and liberation, Illustrated Edition. Harmony, New York, pp 131–133

Heraclitus (2020) Fragments. Digireads.com Publishing, Place of publication not identified, pp 32–38

Marx K (2008) Wage-labour and capital. Wildside Press LLC, Cabin John, MD, p 33

Obama B (2006) The audacity of Hope: thoughts on reclaiming the American dream, 1st edn. Crown, New York, p 243. 261

Rodrigues H (2006) Introducing Hinduism, 1st edn. Routledge, New York, p 51

Rosling H, Rönnlund AR, Rosling O (2020) Factfulness: ten reasons we're wrong about the world– and why things are better than you think, Reprint edition. Flatiron Books, New York, NY, pp 32–38

Silver C, Chen J, Kagen J, Woolsey B (2019) Standard of Living. In: Investopedia. https://www.investopedia.com/terms/s/standard-of-living.asp

Wackernagel M, Schulz NB, Deumling D et al (2002) Tracking the ecological overshoot of the human economy. Proc Natl Acad Sci U S A 99:9266–9271. https://doi.org/10.1073/pnas.142033699

Chapter 2
The Energy Essential: Physical Forces Animate All Things

Abstract Technology purposely channels, concentrates, and transforms energy, matter, and information to improve the human condition. Continuous energy-rich fossil fuel use is the root cause of environmental degradation. Since matter is embodied energy and humankind's demand for *stuff* is at an all-time high, energy sources directly affect future ecological prospects. The proposed solution of shifting to 100% renewable energy is complicated by the need for high embodied energy industrial processes. Understanding abstract concepts of energy is critical to finding solutions for unintended consequences at the root of environmental problems. This chapter demystifies energy terms and establishes conceptual foundations for designers to consider as they work through the primary challenges of consumption.

Technical language and abstract units can obfuscate and seemingly minimize the negative impacts of energy production. Major environmental impacts can hide in dense detail and make it easy for producers to externalize important environmental impacts or shift one ecological cost for another. This chapter provides brief introductions of classical thermodynamic concepts such as *energy*, *exergy*, and *entropy*, as well as the more recent concept of *emergy*. These and the larger underlying scientific phenomena that support them are at the heart of natural and anthropogenic ecological transformations in an industrialized world.

2.1 Design Shapes the Natural World

Any successful next act for *Homo sapiens* will continue to recognize limits and capitalize on opportunities present in the physical world. Those pursuing truly sustainable—or better, regenerative—designs move forward when they simultaneously keep fundamental physical limits and the ultimate goal of the design problem in view. If one main design goal is not only to reduce the harm we do to ecosystems but also to attempt to heal the damage already done with each additional thing we make, we need to first establish reference points in accordance with physical laws that govern all matter and energy.

Most designers work within fairly narrow boundary conditions. They often don't consider the larger physical context in which their designs function. Even so-called sustainable products can externalize fundamental negative impacts by not properly considering or purposely ignoring them as an important part of a system. Scrutinizing

© Springer Nature Switzerland AG 2021

J. Cays, *An Environmental Life Cycle Approach to Design*,
https://doi.org/10.1007/978-3-030-63802-3_2

the larger system in which a design exists early in the process improves long-term well-being.

Consider one of the most pervasive design objects of the twentieth and twenty-first century, the automobile. Representing individual freedom, the ability to go anywhere at any time, the car is responsible for shaping a material percentage of global energy and transportation infrastructure[1] and comprises a significant percentage of global pollution. One proposed solution to the tailpipe emissions problem is to remove the tailpipe and produce only electric cars to satisfy the demand for flexible personal mobility. Still, the current reality requires most cars to draw energy from fossil fuel-based power plants. This means that an electric car that runs on a charge it gets from a coal-powered energy grid is essentially a "coal car."

Burning coal boils water that makes steam in most power plants in the industrialized world. This channeled high-pressure steam spins an electromagnetic turbine to generate electricity. The electricity from this coal-fired plant flows for miles across high-voltage wires and into the electric grid where local substations, power transformers, and distributional transformers (the white or gray buckets at the top of telephone poles) reduce the voltage for domestic use. Only at the endpoint of this massive, mediating infrastructure can this electricity be used to power all the devices in a typical home or office where most electric cars are charged. Conversion losses and embodied energy in the numerous grid and end product components that result from turning one form of "dirty" energy into seemingly "clean" transportation result in net negative environmental impacts.

This coal-powered car is actually worse compared to a gasoline- or diesel-powered car using current technology in most regions. So removing the tailpipe from each car is an illusory fix since there is a much larger and more harmful "tailpipe" at the end of each coal-fired power plant supplying the primary energy source. Without changing the entire grid mix of energy to renewables, one fossil fuel is simply replaced with another. And, from an efficiency standpoint, there is the added loss from intermediary conversions (from electricity production to battery storage) to finally get the car to go.

Solar energy's potential to replace fossil fuel is limited by current technology and infrastructure to efficiently convert, store, and distribute it in the quantities needed to match peak demand. The sun doesn't shine at night, and the wind doesn't always blow. Reliable storage is the solution to all large-scale solar or other energy sources that wax and wane in their availability to produce useful work. We employ numerous techniques with varying efficiency levels in order to smooth out the abundant but undependable flows of solar and solar-induced wind energy available over different times of day or season.

---

[1] According to the National Academies of Sciences, Engineering, and Medicine, in the United States alone, 28% of all energy used goes to moving people and goods. 58% of that is consumed by cars, light trucks, and motorcycles. http://needtoknow.nas.edu/energy/energy-use/transportation/

Our deep dependency on fossil fuels, and the inherent qualities that make them easy to use and hard to replace, compounds the numerous practical challenges to develop clean, high-density, forms of energy. Coal, oil, and natural gas have unique combinations of portability, compactness, high-energy concentration, and relative stability that are well matched with human needs and habits. The easiest to reach fossil fuel stores are already depleted. Challenges in extraction continue to grow, but new technical means and methods have kept pace to increase output of these finite resources.

Without abundant and cheap energy, the supply of ever-new products and services slows or stops. Without this steady supply of goods and services, living standards for the world's top quintile would immediately start to erode. The lower socioeconomic four-fifth would find it impossible to secure the same high standards for themselves without the same energy advantage that the top one-fifth has taken as a matter of course.

Current industrial production and distribution methods, and the consumption behaviors of billions that drive them, cannot continue without carbon-based fuels. The only energy resources left to fuel human activity and satisfy essential and nonessential needs will be those left after we have literally burned through everything else. Setting aside the negative environmental consequences from extracting, processing, and burning them, fossil fuels are finite, and we will eventually run out of them.

So what are our energy options? Any realistic answer to this question from a designer's perspective requires first thinking about what drives our energy systems and the basic forces at play at multiple scales. Science has developed concise language best suited to describe what is going on at all scales of energy conversion. Science is, in fact, primarily concerned with these phenomena. The specialized domains of physics, chemistry, biology, and other scientific fields describe, in minute detail, the interplay between energy and matter and the consequences for human beings. Many who gravitate toward the "creative" design disciplines, however, do so with expectations that they will be engaged solely in nontechnical pursuits. Some suffer through high school and college science and math classes just to complete degree requirements. Motivated by a desire to effect a positive change through sustainable design, they may not yet fully recognize the benefits that even a cursory understanding of basic principles can produce. Our unexamined relationship to energy production, distribution, and consumption continues to cause damage to humans and the environment. Familiarity with some basic terms and concepts from elementary biology, chemistry, and physics covered in this chapter helps designers consider fundamental drivers and limitations. The hope is to expand and improve the range of design responses to reduce future damage and repair damage already done.

2.1.1 Powering the Four Industrial Revolutions

We are now living in the Fourth Industrial Revolution, and while current production methods have materially changed in scale and complexity, they have not changed in kind, since the First Industrial Revolution as all energy conversion technologies follow fundamental discoveries about the principles that govern the natural world. The First Industrial Revolution (eighteenth and nineteenth centuries) was set in motion largely by newly codified thermodynamic laws. These laws, taken together, state that each form of energy at work is a conversion process where nothing is created and nothing is lost but also where *utility* is diminished over time as dense complex forms of energy are transformed to low-grade heat energy unable to affect any more physical changes.

Early twentieth century production methods relied on First Industrial Revolution techniques, such as large scale, high energy, metal refining, and process mechanization, but refocused them through new labor organization systems, like the factory production line, to increase output yields on an unprecedented scale. Fordist assembly line systems were hallmarks of the Second Industrial Revolution, which drove a globalizing economy with increasing efficiency until well into the third quarter of the last century. Large features of this Second Industrial Revolution persist today as traditional factory output of tangible goods continues to increase in absolute amounts. Information technology started reshaping the flows of matter and energy on a small scale in the 1950s and grew to become a dominant force in the Third Industrial Revolution, generally referred to as the "Information Age." Networked computer technology tracks, records, and stores massive amounts of data as products in themselves as well information about things and their relationships within complex systems. Each Revolution has built upon, rather than fully supplanted, the previous one. Each subsequent development depends on the infrastructure laid down by previous ages. The Fourth Industrial Revolution is further synthesizing developments of the first three and reshaping global economic activity in the process (Schwab 2017). A key feature of this latest turn is the simultaneous increase of energy, data, and material flow, to produce the abundance of goods and services, as well as the inclusion of sensors and transmitters designed to link all of it together as the physical and informational merge in the Internet of Things.

From First to Fourth, scientific discoveries and technological advances drove massive growth in industrial as well as agricultural production, and all share in simultaneously increasing the standard of living and life expectancy for more and more people on the planet.[2] These increases would be impossible without a constant increase in the conversion of energy from carbon-based fuel and the concomitant negative environmental impacts that follow. The thermodynamic principles that explain this relationship have been clearly understood for centuries. They are

---

[2] The early twentieth century Haber-Bosch process, for example, produces ammonia-based fertilizer by heating air to over 700 degrees F and pressurizing it to over 2000 psi to extract nitrogen that is readily available for plants (Smil 2008).

consistent at multiple temporal and spatial scales and can, therefore, be expected to act in predictable ways to explain many important effects of human industrial activities.

2.2 What Is Energy?

Since the nineteenth century, we have described energy, indirectly, as units of work. In his lecture on the Conservation of Energy, the Nobel laureate, Richard Feynman states as a matter of fact that: "we have no knowledge of what energy is. We do not have a picture that energy comes in little blobs of a definite amount" (Feynman 1963).

Energy is a word that describes the potential of abstract physical forces to transform a system. The language scientists and engineers use to discuss its practical effects deals largely with an accepted understanding of the relationship between units of force, energy, and power. The work that is accomplished takes place in space over time. Wilhelm Ostwald's Fundamentals of General Energetics begins by identifying energy as a phenomenon as fundamental as that of time and space:

> The concepts that find application in all branches of science involving measurements are space, time, and energy. The significance of the first two has been accepted without question since the time of Kant. That energy deserves a place beside them follows from the fact that because of the laws of its transformation and its quantitative conservation it makes possible a measurable relation between all domains of natural phenomena. Its exclusive right to rank along space and time is founded on the fact that, besides energy, no other general concept finds application in all domains of science.
>
> Whereas we look upon time as unconditionally flowing and space as unconditionally at rest, we find energy appearing in both states. In the last analysis everything that happens is nothing but changes in energy…Experience shows that the decrease of energy at one place is always associated with an equal increase somewhere else. Ostwald (1892 trans. Lindsay 1976b)

Essentially, if Cartesian space is 3D, and the whole of the indefinite progression of existence and events from past to present and into the future is 4D, then energy can be considered as is an AD or "all dimensional" phenomenon.

2.2.1 More Precise Energy Terms: Exergy, Anergy, and Entropy

Understanding what happens when one form of energy is converted into another or when it is temporarily stored away as matter requires a broader vocabulary beyond the word energy.

Exergy – The amount of available energy to do work. High exergy systems include the fusion reactions in our sun and fossil fuels that captured, concentrated, and stored electromagnetic radiation that flowed from it millions to hundreds of thousands of years ago. In statistical terms exergy is "the maximum fraction of an energy form which (in a reversible process) can be transformed into work."

Anergy – The remaining part, corresponding to waste heat, is called anergy (Honerkamp 2002)—the two, exergy (from the Greek ex + ergon for "from work") and anergy[3] (from the Greek an + ergon for "without or no work"), make up the two sides of the energy balance sheet. This holds for any local analysis of a discreet or closed system.

In all irreversible systems, low-quality, anergetic, waste heat increases as high-quality exergy decreases. The trend in all isolated systems, no matter how small or large, follows this one-way progression. Without an outside exergy source to contribute to the plus side of the balance sheet, the system will continue to transform high-quality types of energy, able to perform work, into disorganized, low-quality types of energy unable to do work.

Entropy – This transformation is called entropy. Rudolph Clausius introduced the term in 1865 as a counterpart to energy replacing the Greek root ἔργον (érgon, "work") with τροπή (tropé, "transformation").

> I hold it to be better to borrow terms for important magnitudes from ancient languages, so that they may be adopted unchanged in all modern languages, I propose to call the magnitude S the entropy of the body, from the Greek word τροπὴ, transformation. I have intentionally formed the word entropy so as to be as similar as possible to the word energy; for the two magnitudes to be denoted by these words are so nearly allied in their physical meanings, that a certain similarity in design appears to be desirable. Clausius (1867)

Nineteenth-century thermodynamic energy concepts were key in describing how both energy could be transformed by both natural and mechanical processes. These developments in knowledge about thermal energetic systems were part of a much larger scientific system of knowledge that had been growing for centuries and continues now.

2.3   Four Primary Forces

All change is fundamentally made possible and limited by forces constituting all of nature. At the start of the sixteenth century, Nicolaus Copernicus started a 500-year-long scientific revolution that continues to this day when he described solar-centered planetary motion for the first time.[4] Using better tools and developing more precise mathematical models, Johannes Kepler in 1609 (Donahue trans. 1992) and Galileo

---

[3] Anergy is not a classical thermodynamics term. Coined by the Head of Strategic Planning of Shell International Petroleum, Guy Jillings, in the 1980s, it has since been adopted by both statistical physics and immunobiology. It is a useful direct counterpart to exergy.

[4] This scientific and technological revolution is founded on his observations of planetary revolutions that he first distributed in his *Little Commentary* (1514). A few decades later he published mathematical descriptions of planetary motion in *De revolutionibus orbium coelestium* (*On the Revolutions of the Heavenly Spheres* 1543).

Galilei in 1632 (Galilei and Favaro 2020) refined the theory of gravitational influence in the century that followed. Patient observation, new tools, and mathematical interpretation revolutionized cosmology. We were no longer at the center of the universe. Understanding how the invisible forces of the sun's gravity shape the elliptical paths of the planets under its influence opened the door for human beings to understand and harness these essential physical forces to systematically shape the modern world.

Gravity is one of the four primary physical forces that combine to govern how the material universe behaves. The three other forces that work directly on atomic and subatomic levels are electromagnetic (Maxwell 2010), nuclear strong (Chadwick 1932), and nuclear weak forces (Fermi 1934). The history of technology over the last millennium traces our ability to perceive and direct these forces to accomplish increasingly complex and ambitious tasks.

While *force* is distinct from *energy*, the dynamic interplay and conversions between electrical, chemical, mechanical, thermal, and other forms of energy, which issue from these forces, set into motion everything on earth. The detailed study of these phenomena and their effects are the concerns of entire scientific domains and lie beyond the scope of this book. However, in order to realistically consider the threats and opportunities we all face, it is necessary to keep in mind that all matter and energy behave according to fundamental physical laws.

2.3.1   Electrochemical Forces

From the beginning of time and space, these forces unceasingly transform material systems at every scale from the galactic to subatomic. We are, however, primarily concerned with the practical problem of how energy affects us, and the things around us, today and into the relatively near future. The vast majority of stored chemical energy available on earth is transformed from one single fundamental source, the sun. Organisms such as plants, algae, and some bacteria that synthesize their own food from inorganic substances and an external energy source are called *autotrophs*. Most autotrophs are photoautotrophic plants that convert the electromagnetic energy in sunlight directly into burnable chemical energy stored in the plant's cells. In addition to sunlight, they take in water, mineral nutrients, and atmospheric carbon dioxide and turn it into sugars they use for food through photosynthesis.

Photosynthesis is the single most important energy transformational chemical process for nearly every species on earth. It sustains all life. From a point of view that seeks maximum efficiency in solar energy conversion, photosynthesis is not very *efficient* at all. On average, less than 1% of the sunlight is converted into chemical energy through photosynthesis. Given the abundance of plant life on earth, however, maximum efficiency is not a long-term goal for the *effective* use of solar energy for all photoautotrophs. The oxygen they "exhale" as a waste byproduct and the chemical energy they gather, change, and store in their tissues directly and indirectly support all animal life on earth.

The energy stored in renewable plant-based biofuels (wood, charcoal, peat, etc.) has been civilization's energetic limiting factor for millennia. (This changed as Industrial Revolution mining practices created access to ancient underground stores of concentrated fossil fuels.) *Heterotrophs*, including all animals, eat plants and convert that chemical energy to power their own bodies' metabolisms and locomotion. A continuous supply of nutritious food creates an energy web where innumerable different organisms eat and are eaten. The primary solar energy passing through this web continues to sustain the majority of life of earth. Almost everything produced by humans during the last 10,000 years is a channeling of excess primary solar energy, a surplus available beyond what is needed to survive. Whether through the directed efforts of beasts of burden or magnifying the effects of the relatively limited human strength through simple machines, human settlements were sustained for millennia by agrarian economies (Smil 2008). Today's nearly 8 billion people, living in the midst of the Fourth Industrial Revolution, continue to be utterly dependent on our sun for survival. And as long as humankind inhabits the earth as we will take advantage of the "almost inconceivably minute fraction [a one hundred and forty thousand millionth] of the Sun's heat and light reaching the earth…the source of energy from which all the mechanical actions of organic life, and nearly every motion of inorganic nature at its surface, are derived…" Kelvin (1854 trans. Lindsay 1976a).

Although each new scientific discovery reveals deeper, more precise, and more sophisticated insights into nature's workings and the constant interplay of its forces at work, our day-to-day lived reality continues to largely reinforce many of the early findings of theoretical and empirical scientists working since the scientific revolution began.

2.4 Sharing Knowledge Accelerates Energy-Intensive Inventions

Individual contributions from Copernicus through Newton laid the foundations for modern science. This Scientific Revolution reached a critical point in the mid-seventeenth century when discoveries of new provable natural laws moved from publication in expensive and relatively rare books to their systematic collection and dissemination available to wider audiences. In 1660 Charles II granted a royal charter to a collective of "natural philosophers" working across diverse scientific domains to establish the Royal Society of London for Improving Natural Knowledge. The Royal Society published the first peer-reviewed scientific journal in 1665. It remains the oldest scientific journal still in publication.

The dissemination and free exchange of reproducible experiments and verified facts, since then, have resulted in a steadily growing number of trusted, specialized, peer-reviewed scientific journals and apply industrial practice publications. The practical outcome of a deeper understanding of fundamental physical and chemical laws governing energetic systems and the sharing of these discoveries were ever higher work yields through new engines of industry.

Less than a century after Isaac Newton's *Principia* (1687) laid the mathematical capstone for modern science,[5] (Motte, trans. 2018) British industrialists began applying both abstract and empirically tested scientific principles, especially in the emerging field of thermodynamics, to maximize work output.

Between 1763 and 1775, James Watt made continuous improvements to Thomas Newcomen's 1712 steam engine. His 1769 patent, for a *New Method of Lessening the Consumption of Steam and Fuel in Fire Engines,* marked a more than doubling in power output (Smil 2008). Eighteenth-century English mining, iron production, and textile manufacture were the first industrial processes supercharged by these new engines. With relatively portable and continuous steam energy sources, industry could establish itself far away from high-power mill wheels driven by large, fast-running rivers and waterfalls. The First Industrial Revolution that was set in motion by new knowledge and invention replaced human- and animal-powered work, renewable biomass, and renewable energy-sourced (wind and water) power plants with fuel sources containing higher specific energy. Ever-expanding industrial production from that point until today was and is primarily powered by fossil fuels. Moreover, what is produced has changed to include goods and services that, in the aggregate, require enormous amounts of fossil energy to function.

2.5 Sadi Carnot's Caloric Thermal Energy

The familiar calorie unit measures the energy content in foods. The word itself comes from "calor" the Latin word for "heat"; it directly names energy as heat. Today's nutritional unit is the kilocalorie also referred to as the large calorie and is equal to 1000 small calories. 1 small calorie is the amount of heat energy needed to raise the temperature of 1 gram of water by 1 degree centigrade.

In Sadi Carnot's self-published book, *Reflections on the Motive Power of Heat and on Machines* 1824, he described an ideal heat engine capable of converting 100% of heat within a system into mechanical motion that he called "motive power." From studying the way existing steam engines worked,[6] Carnot (1897) extrapolated a theoretical maximum efficiency where no loss of motive power was possible under the right conditions. He relied on Antoine Lavoisier's widely held but erroneous notion of a ubiquitous invisible fluid called "caloric" that flows in only one direction from hotter to colder bodies. According to Carnot, this fluid can more efficiently

---

[5] Prior to this landmark work, Gottfried Leibniz published his paper, "A New Method for Maxima and Minima" in 1684 and is credited for developing calculus in parallel with Newton.

[6] Carnot acknowledges the practical and incremental contributions that produced ever more efficient engines by Thomas Newcomen, James Watt, and other English engineers in the opening pages of his seminal work.

perform mechanical work as the difference between the hotter reservoir and colder reservoir grows larger.

Carnot's unpublished notes from a few years later contain ideas that attempt to directly connect heat to indestructible energy or motive power:

> Heat is simply motive power, or rather motion which has changed form. It is a movement among the particles of bodies. Wherever there is destruction of motive power there is, at the same time, production of heat in quantity exactly proportional to the quantity of motive power destroyed. Reciprocally, wherever there is destruction of heat, there is production of motive power. We can then establish the general proposition that motive power is, in quantity, invariable in nature; that it is, correctly speaking, never either produced or destroyed. It is true that it changes form, that is, it produces sometimes one sort of motion, sometimes another, but it is never annihilated. Carnot (1830 Trans. Thurston, R.H. 1897)

Twenty years later, this insight into the conservation of energy would be codified by Rudolph Clausius as the first of four thermodynamics laws.

2.5.1   First Three Laws of Thermodynamics

Three laws of thermodynamics were formalized during the last half of the nineteenth and at the very start of the twentieth centuries. They establish energy's eternality and indestructability, its one directional flow through time, and the limit of any system to be completely devoid of it.

The First Law of Thermodynamics states that energy cannot be created nor destroyed. It only transforms from one type of energy to another. Formalizing the earlier work of Carnot, Clausius succinctly posits that "In all cases in which work is produced by the agency of heat, a quantity of heat is consumed which is proportional to the work done; and conversely, by the expenditure of an equal quantity of work an equal quantity of heat is produced" (Clausius, 1850, in Ganesan 2018).

The Second Law holds that energy's capacity to do work diminishes through time. Clausius goes beyond the work of Carnot to explain the inexorable movement of everything toward a state of disorder. He writes that "heat can never pass from a colder to a warmer body without some other change, connected therewith, occurring at the same time" (Clausius 1867). All matter and well-ordered concentrations of energy, therefore, tends to dissipate into less organized states or lower-quality energy (heat) and toward maximum entropy.

The Third Law establishes that entropy can only approach zero degrees. Walther Nernst heat theorem in 1906 establishes the practical fact that the universe or any open or closed system within it can never reach absolute zero.

> It is impossible for any procedure to lead to the isotherm $T = 0$ in a finite number of steps (Bailyn 1994).

2.5.2   Zeroth Law of Thermodynamics

Many other things are exceedingly difficult to measure as well. Qualitative or ephemeral experiences, for instance, defy direct and precise quantification. Psychological states, aesthetic effects, and changing public opinion, for instance, can only be described or approximated with heavily qualified statements. Temperature, however, is measurable once a scale is agreed upon and standardized instruments are properly calibrated.

The Zeroth Law of Thermodynamics was formalized and labeled in 1935 by the British physicist Ralph Fowler after the first three laws were already formulated. It definitively establishes the transitive relation (if A = B and B=C, then A = C) between thermal systems in order to measure heat. It says that if two thermodynamic systems are each in thermal equilibrium with a third, then they are in equilibrium with each other.

This is important because temperature is an abstract notion. Humans are well adapted to survive in a thermally active or variable environment. Thermoreceptors in our skin respond to "hotter than" and "colder than" stimuli as subjective sensations. Our innate physiological grasp of thermal differences, though, cannot provide a common notion of temperature. Beyond the general subjective and qualitative nature of human sense perception of thermal energy, sensitivity varies between individuals, and even depends on which part of the body does the sensing, and it generally decreases with age (Stevens and Choo 1998).

Thermometers were invented to express the amount of heat as a quantitative, numerical representation. The Zeroth Law recognizes the fundamental need to use and trust instruments to provide a common understanding of the precise amount of heat in an object or system. In light of today's challenges to trust stated scientific facts created in a cultural environment where everything is relative and open to interpretation, it may help to generalize and extend the Zeroth Law to other instruments of measure where applicable and where enough equivalencies are available to establish a standard for calibration.

Agreement between various instruments displaying environmental impact information is discussed in Chaps. 7, 8, and 9. In contrast to the complexities inherent in compound, multistage, and multi-metric analysis such as life cycle assessment, temperature is a relatively straightforward single indicator. Temperature, nevertheless, requires an agreed upon faith in the accuracy, precision, and overall veracity of instrumental readout. Most importantly, it depends on the significance and utility of the information gleaned every time a thermometer is consulted (Fig. 2.1).

When the mercury in a thermometer stops rising or falling, it has achieved thermal equilibrium with the substance it is measuring. If the reading doesn't change from one thermodynamic system to another, it is accepted that they are each in thermal equilibrium with the other. It is so commonplace today that few would now question the once abstract concept of temperature. Even the most ardent science denier acquiesces to the Zeroth Law at least tacitly and doesn't question the underlying faith we each place in measuring the rate at which molecules are moving in a substance in contact with the surface of the thermometer.

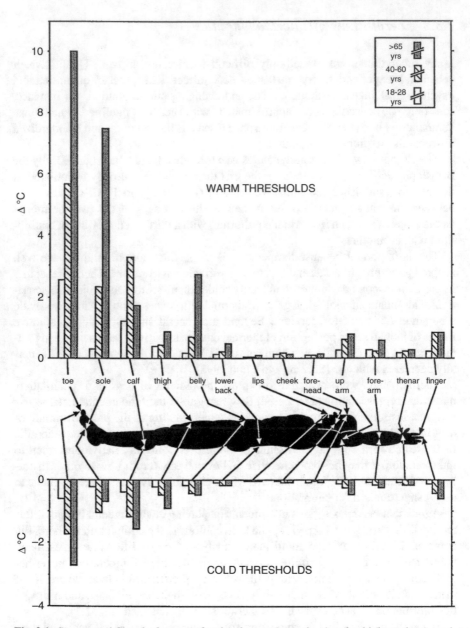

Fig. 2.1 Stevens and Choo body maps of regional warm (upper bars) and cold (lower bars) sensitivity over a typical life span. The bars indicate the median thresholds of 3 groups of 20 young, 20 middle-aged, and 20 elderly subjects. (Licensed for editorial use by Taylor & Francis)

   The four laws of thermodynamics taken together help us to think about energy exchanges in terms of one-way transformations of highly concentrated, well-ordered, sources into dissipated lower-grade heat energy that is no longer capable of effecting change through doing work. They also establish the general boundary conditions inside of which we all live and between which we can measure changes. The relative harm or good these measurable changes cause depends on spatiotemporal scales being evaluated. Measurable local short-term gains, made by harnessing energy transformations through technological advances, may yield long-term harm elsewhere or, in the case of the environment, to the overall larger system.

2.6   Joule: The Official Energy Unit

James Prescott Joule was a nineteenth-century British physicist who explored the relationship between heat and work. In a series of experiments carried out between 1845 and 1847, Joule used three custom-built instruments to precisely measure the subtle temperature rise from the friction caused by gravity-driven small paddles rotating through a vessel filled with water. The "Joule apparatus" experiments demonstrated the direct mechanical transfer of diminishing gravitational potential energy as a string connected to a descending weight spun the bearing mounted in the brass paddle shaft to proportionally increase internal heat energy in the water. This early experimental proof of the conservation of energy was, according to Joule, carried out in "a spacious cellar, which had the advantage of possessing an uniformity of temperature" (Joule 1850), and recorded the results to 1/200 of a degree Fahrenheit. Although there were others working at similar purposes, Joule was the first to carefully control and document his experiments linking two types of energy.

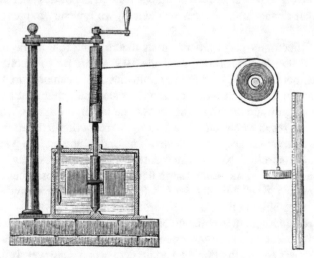

The Joule apparatus originally published in Harper's New Monthly Magazine, No. 231, August, 1869. (Image in public domain)

For his contributions to the field of thermodynamics, the International Standard base unit of energy in any form is the "Joule." One human-scaled example of the amount of energy measured in a single Joule is the work done by lifting an apple (with a mass of a bit over 100 grams) off the floor to a height of 1 meter. A less relatable example connecting thermodynamics to electrodynamics is 1 joule of heat that is given off each second as 1 ampere of electrical current flows through 1 ohm of resistance. For many, unfamiliar with electrical units, this requires clarification and further definition of terms.

2.6.1 Power Is Flowing Energy Measured Over Time

Power is often confused with energy. Power is distinct from energy in that it is measured over time. Energy is an amount, and power is the rate at which this quantity is converted from one type of energy to another. A watt is a unit of power. Whether referring to electrons, wind, or steam, the flow of electrical, mechanical, chemical, or thermodynamic energy can all be expressed in watts. A year before his death, William Siemens proposed in 1882 that the unit be named after the Scottish inventor of the steam engine, James Watt, who a hundred years earlier "first had a clear physical conception of power, and gave a rational method of measuring it" (Siemens 1883). Watt recorded the mechanical power output of a mill horse of 32,400 foot pounds per minute. Today watt is used by most to describe electrical power such as how much a lightbulb consumes to generate illumination.

Any discussion of the power output or consumption of something, whether it is a hydroelectric dam or a 60 watt lightbulb, must refer to a duration even if not explicitly stated. Seconds, hours, days, years, or some other time unit can be used, but when referring to power ratings of things as large as power generation plants to relatively small domestic appliances or devices, we typically express power flow in hours.

Harnessed electricity plays an enormous macro scale role in the industrialized world. The 23,000 terawatts/hour used in 2019 according to Enerdata's Global Statistical Yearbook 2020 (World Energy Statistics 2020) can obscure the ceaseless action of electron transfer at the microscopic scale in all chemical and biological activities. From galvanic reactions that cause one metal to dissolve when touching another that can result in structural failure to the reduction-oxidation reactions of burning fuel, these are examples of electrons naturally moving in a way that can be anticipated and controlled. Net environmental damage results largely through energetic reactions at this atomic scale. Large flows of harmful substances released into the soil, water, and air chemically react to form persistent compounds, not readily metabolized, destabilizing the earth's natural systems.

Whether considering environmental electrochemistry, industrial-scale electricity production, or consumer-scale power usage and needs, the terminology is the same. One watt of power equals the flow of 1 joule per second. One watt is the force of 1 newton (the same apple pressing down on the hand lifting it) pressing against an object moving at a constant velocity at 1 meter per second. When discussing electrical power, other base unit terms collectively (and self-referentially) describe the various aspects of the electron flow. Voltage is the potential energy difference between two points. Named after Alessandro Volta, the inventor of the battery in the last decade of the eighteenth century, the volt is the unit of electromotive force. His published experiments demonstrated how one metal willingly gives up its electrons to another in the presence of a liquid electrolyte to produce an electric current.

Amperes, named after the father of electrodynamics, André-Marie Ampère, quantifies electrical current of an electrical charge. One ampere equals 1 coulomb (this is a group of approximately 6.242×1018 electrons) of electrical charge passing a given point each second. Conversely 1 coulomb is the amount of electrical charge carried each second by 1 ampere of current.

Ohm is a measure of electrical resistance. Published in 1827 by George Ohm, Ohm's law states that the current through a conductor between two points is directly proportional to the voltage across the two points. The resistance created is proportional to the current through the conductor in units of amperes which equals the voltage measured across the conductor in units of volts, over the resistance of the conductor in units of ohms. As the current or amperage increases, the resistance decreases.

Electrical potential and flow can be conceptually related to water flowing from a pressurized tank. The volume of water in the tank is similar to the electrical charge measured in coulombs. A larger tank holds more than a smaller one. The pressure in the closed tank determines the potential the water has to act on something outside the tank. This is the water tank corollary of the difference in charge between two points measured in volts. The higher the pressure, the more potential the water (or the electrons) has to cause an effect.

Why is this important to design? More than the specific technical measurements and units describing the effects of work done in a system, it is most important to understand that all types of energy are interchangeable and equivalent with respect to the work they produce and the heat byproduct created after that work is completed. The total amount of matter and energy is neither subtracted nor added in the process. Even though nothing is lost nor gained quantitatively, much can change qualitatively affecting the conditions on which living creatures rely.

Incremental environmental effects wrought by industrial activities that assemble so many new physical and chemical compounds are, at times, pronounced, but most accumulate silently over longer periods too subtle to readily discern. Many energetic reactions do no harm at all over time, but some, like those that release carbon dioxide, sulfur dioxide, or nitrogen compounds into the ecosphere, are unequivocally linked to destabilizing environmental dynamics. Designers who consider these basic principles are better positioned to provide alternative solutions to satisfy needs without harmful effects.

2.7   Chemical Balance

An imbalance of some chemical compounds can lower the environment's ability to support life by changing the conditions under which life has developed over millennia. Most life forms cannot survive outside a narrow range of temperature, chemical balance, aridity, nutrient intake, and other parameters that make homeostasis possible. It is easy to implicate certain elements or compounds in causing all current ills of the world. Too much carbon in the atmosphere has been directly linked to climate change, but eliminating it altogether would lead to the collapse of all plant life on earth. Even cursory considerations show that no element or compound is inherently bad (or good) in every case. Rather, only when they exceed certain concentrations do they cause harm.

This happened 2.35 billion years ago when cyanobacteria began photosynthesizing and oxygen levels began to rise, eventually triggering a massive die-off of anaerobic microbes. The dominant life forms could not mount any defense against the deadly toxic effects of an oxygenating atmosphere, and life on earth was nearly extinguished (Blaustein 2016). The same result from the more recent end-Permian mass extinction event caused by large-scale volcanism in Siberia 251 million years ago where all life on earth was nearly wiped out. Again, this was due to a large change in the air and water to which species could not adapt. Shifts in atmospheric chemical composition transfer to ocean chemical composition led to the most devastating ecological event of all time (Sahney and Benton 2008). Anthropogenic changes to the chemistry of our air and water may again tip the balance and contribute to the next mass extinction.

Energetic chemical bonds reacting between relatively small numbers of elements are responsible for most natural resources on which we depend. We have found industrial uses for most of the 118 elements on the periodic table including the synthesized 30 heaviest radioactive ones used in atomic research. Each compound emerges through processes that aggregate molecules in special organic and inorganic configurations. Metal alloys, wood, stone, petroleum, soil, water, food, and breathable air are all examples of the chemical compounds we depend on. The 10 most ubiquitous elements in the human body are found among the first 20 lightest ones listed. These same elements that combine to form over 99% of our bodies are found everywhere else in the natural and built environment. Take out sodium, and they are the same macronutrients found in rich and productive soil (Lindbo et al. 2012), supporting the adage that "we are what we eat" and demonstrating the staggering complexity that can arise through the combination of a small subset of elemental building blocks. Each arrangement defines their bond properties and how much energy it takes to create, sustain, or break them.

Oxygen, the third most abundant element in the universe, is a powerful accelerant that drives combustion and other one-way oxidation reactions causing a substance to give up its electrons atom by atom to oxygen. Steel rusts when oxygen reacts with iron and electrolytes present even in moist air. Fire burns when oxygen reacts with a fuel source (often carbon based) and heat. Carbon not only oxidizes when organic matter ignites and burns with a plentiful supply of air but also during animal cellular respiration. All these processes are irreversible.

2.8   Work and Value

The seventeenth century ushered in the Enlightenment and accelerated new techno-logical developments throughout every subsequent century. As newly specialized scientific disciplines, primarily biology, chemistry, and physics, produced empiri-cally measurable definitions for work, economic systems began to value maximiz-ing productive yields on energy investments. High-value goods and services typically require high oxidation rates of carbon in one form or another. In an age before machines burned fossil fuels, energy was primarily converted and embodied through human or animal labor.[7] More work required more physical effort, and there was a biological limit to what could be accomplished.

2.8.1   Animal Power

Work produced by the bodies of people and draft animals is, at root, a product of the chemical energy conversion by adenosine triphosphate (ATP) to adenosine diphos-phate (ADP) through hydrolysis (as ATP reacts with water) in the cellular metabo-lism process. This is the fundamental energetic mechanism of all life. It works like a toggle switch. Each time a phosphate is loosened from one ATP molecule in a cell, tiny amounts of energy are released.[8] The macronutrients (carbohydrates, fats, and proteins) and micronutrients (vitamins and minerals) in food supply new high-energy electrons to toggle the switch back, and ADP is converted into ATP again. This extremely small amount of energy is multiplied by a staggeringly large number of molecules working constantly. Each of our cells contains approximately a billion ATP molecules. This flow of electrons powers all biosynthesis processes, the main-tenance of cellular homeostasis, and the contraction of muscles. Oxygen is the most common element in the human body. Proper metabolism in animals is possible only when oxygen is available as a catalyst as discussed below and even though oxidative phosphorylation (ADP to ATP) acts as a reversible switch, electron leaks create

---

[7] Notable exceptions were sailing ships propelled by wind and mills that employed the kinetic energy of wind and water to operate when these renewable energy resources were flowing (Smil 2008, p. 201).

[8] About 31,000 joules per mol of ATP molecules are released in humans at normal body tempera-tures (Atkins and Paula, 2010, p. 212). Some mammals may release up to 50 kj per mol of ATP (Smil 2008, p. 39).

reactive oxygen species. These byproducts produce free radicals that, over time, cause irreversible cell damage aging and disease (Valko et al. 2006). All elemental oxidation is caused by electrons of one element moving and combining with another. The resulting energy released overtime is responsible for the growth and sustenance, as well as contributing to the ultimate destruction, of living organisms.

Oxygen in the air we breathe acts as a catalyst to supercharge the breaking of the phosphate group at the end of the ATP molecule. In addition to producing energy and water, cellular respiration also produces CO_2 that each cell expels. The body collects this byproduct in the bloodstream, carries it to the lungs to be exhaled. Plants directly convert and store solar energy, and photosynthesis drives plant growth through a similar ATP/ADP cycle as takes place in animals. The O_2 byproducts generated through plant cell respiration are expelled into the air in tremendous quantities ready to be used again by the animal kingdom in a tightly coupled exchange of carbon dioxide and oxygen (Falkowski and Isozaki 2008).

In addition to the clear benefits an oxygen-rich atmosphere provides, people and animals also eat oxygen-producing plants to build muscle, regulate electrochemical balance, and maintain homeostasis. Each day, we convert more than our body weight of ATP to accomplish all of these essential life functions. In a 75-year life span, a typical 70 kg (155 lb.) person will generate approximately two million kg (4.4 million lbs.) of ATP out of ADP and inorganic phosphate (Senior et al. 2002). The net effect from the flow of electrons is that animals, including humans, are able to convert energy in the food and water they eat and drink into the energy needed to maintain minimal bodily functions. Any surplus, beyond what is needed to simply keep the organism alive, can be channeled into action, outwardly directed into the surrounding environment. All these activities, from the essential needed to survive to the surplus available to affect other changes in the larger world, are *work* in the broadest sense of the word.[9]

Preindustrial societies also assigned greater value to human and natural products that concentrated energy in the form of human or animal labor. From purely artistic works, to utilitarian objects and facilities, to agricultural stores of nutritious food, people appreciated, coveted, and fought over goods containing higher energy concentrations. The advent of new industrial technologies and processes would eventually replace the relatively low and slow energy flows used in animal-powered processes for millennia. New processes could concentrate unprecedented amounts of energy made possible by the burgeoning applied sciences.

---

[9] The original Greek word for work – ἔργον (*ergon*) is added to the word for in – εν (*en*) to form the compound ενέργεια (*energeia*) from which the word *energy* is derived (Smil 2008).

2.9   Burning Fossil Fuels Releases Stored Carbon-Based Energy

Carbon when combined with oxygen and a source of ignition produces a strong exothermic reaction. It releases a lot of energy. Coal is approximately 50% pure carbon. Carbon-based life itself is possible only through the hydrogenation of carbon dioxide. Organic chemistry is separate from inorganic chemistry in that it studies organic molecules that always contain carbon and usually hydrogen. Inorganic chemistry is concerned with inorganic reactions and molecules that usually don't contain carbon.

Since all life on earth is carbon based, it is impossible to support wholesale vilification of this element. That said, higher levels of atmospheric inorganic carbon compounds especially carbon dioxide (CO_2) and methane (CH_4) are identified as greenhouse gasses and implicated as climate change drivers.

High-energy industrial processes, such as the manufacturing of steel, concrete, and paper, the refining of fuels, the production of fertilizers and other chemicals, and large-scale electricity generation to power the gird, require a steady supply of high exergy fuel. Each fuel has a *specific energy*, which is the amount of energy embodied in the *mass* of a substance; it is measured in megajoules per kilograms (MJ/kg). A fuel's *energy density* is the amount of energy contained in a bounded *volume* of a substance, such as a tank holding a fuel, measured in megajoules per MJ/kg.

Figure 2.2 shows the comparison of both specific energy and energy density for a range of different fuel types. It is clear from this graphic comparison why coal, as well as liquid hydrocarbons, drives modern civilization. The relative convenience, transportability, transformability, and safety that different forms of energy provide are key drivers of any fuel's utility and popularity. Uranium 235, for instance, boasts specific energy of over 83,000,000 MJ/kg, an energy density of 7.4295E+10 MJ/m^3, and zero CO_2 emissions. There are significant associated process complexity involved in splitting uranium atoms to yield energy, high facility costs, and problems with radioactive waste disposal afterward. Uranium is, therefore, not as widely used as high carbon content coal with its comparative paltry specific energy six orders of magnitude lower, at 30 MJ/kg, and its high CO_2 emissions.

2.10   Matter Is Embodied Energy

For the designer, the matter/energy connection is significant beyond the fact that it takes energy to manufacture goods. At the smallest elemental scale, energy is constantly being exchanged and transformed. This fact makes "net zero energy" products, in the strictest sense, at once impossible and unavoidable. Whether any good or service can be considered net zero depends on the boundaries that define the accounting and the scale of what is being measured. This is useful to designers

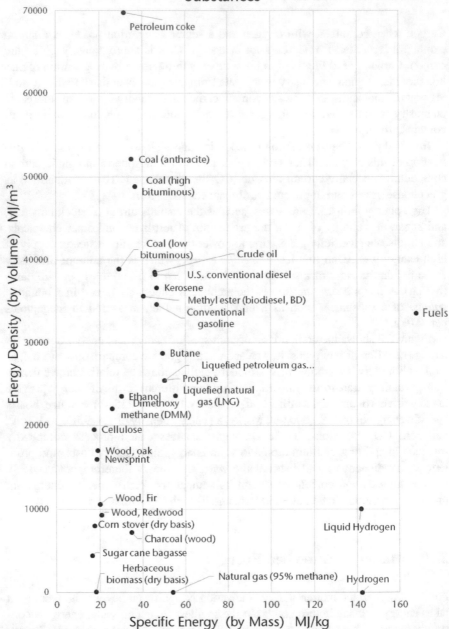

Fig. 2.2 Graphically comparing various energy sources helps to see how energy-dense coal, in its compact and convenient, became a dominant energy source throughout the nineteenth and into the twenty-first centuries. Fuels higher up on the graph concentrate more energy in a smaller volume. (Primary data source: engineerstoolbox.com)

concerned with energy conservation in their designs. It is becoming a more widely accepted knowledge especially in building design.

Keil Moe and William Braham are two architects and academics, for instance, who utilize and teach a comprehensive environmental accounting systems approach to guide primary architectural design considerations. Both carefully contextualize buildings as open thermodynamic systems able to be better designed to harness the free energy that flows continuously into, out of, and around them. Keil Moe writes extensively on the twentieth-century concept, *emergy*, or "energy memory," to attempt to capture energy's dynamic play over time through a range of building materials (Srinivasan and Moe 2015). Emergy was a term coined by systems ecologist Howard T. Odum and is useful when attempting to track both quantitative and qualitative energy transformations throughout systems at multiple scales (Odum 1995).

Although the boundaries of emergy analysis provide the fullest possible measurement of the embodied energy of labor, environmental systems, and solar energy (GAO 1982) of all currently available assessment methods, accurately tracking, measuring, and calculating emergy values are extremely challenging.

The chemical byproducts of high-quality energy released from fossil fuels lie at the root of all environmental degradation problems. If the sources from which we extract energy are the primary drivers of climate change, then common sense dictates that once the shift to a renewable energy powered world is complete, the problem is solved. We could produce and use as many things as we would like without any negative consequences. The fact is, however, that while decarbonizing the industrial economy is a rational and worthwhile endeavor, there are significant technical hurdles to overcome. As long as we need to refine raw materials in bulk using high-temperature industrial processes, the need to burn energy-dense materials will persist. While current industrial innovations continue to improve our prospects for powering heavy industry with low-carbon processes and alternative energy, we may be shifting ecological burdens to other areas. Furthermore, any strides made to marginally reduce harmful effects may lead to even larger negative consequences over time.

2.10.1   The Jevons Paradox

Attempts to blunt the worst environmental impacts of energy consumption through increasing system efficiency are limited when overall demand continues to outpace incremental improvements. Eighteenth- and nineteenth-century scientific discoveries precede and coincide with the accelerating rise in global temperature throughout the twentieth century. The Jevons paradox, first postulated by the British economist, William Stanley Jevons in 1865, posits that *successfully striving to improve energy efficiencies of any device or system to reduce the use of primary fuel sources leads ultimately to increasing consumption*. This is due to a rebound effect caused by greater demand by more people who want to benefit from the improved system. Jevons concludes that: "It is wholly a confusion of ideas to suppose that the economical use of fuel is equivalent to diminished consumption. The very contrary is the truth" (Jevons 1865). This point has been debated since he first published it but highlights the importance of considering what it takes to create and support any individual system (product or process) in the context of the larger ones in which it operates. This means that all the incremental efficiency improvements successfully designed into goods and services intended to reduce overall energy consumption will likely fail at the largest scale.

2.11   Design's Role

Design is, essentially, a "negentropic" act in a universe that tends toward dissolution, homogeneity, and chaos. Design as a process, along with its products, channels exergy to create concentrated local pockets of order and complexity. These activities run in parallel with the multitude of natural processes doing the same thing as they harness the enormous solar primary energy supply that supports all living beings. In the absence of a continuous source of exergy, nature moves to form more stable chemical and physical structures using the same elements. These atomic scale arrangements that require less exergy to maintain allow the constituent elements to assume an equilibrium with one another and with their environment. It takes added external thermal, chemical, electrical, or other kind of exergy acting on a self-organized stable compound to break it back down into its constituent parts.

For twenty-first-century designers and consumers, there is an enormous amount of solar "free energy" to harness. This super abundant electromagnetic energy reaches us daily. It is stored in the living tissues of plants and plant-eating animals including us. The remains of ancient organisms carried this solar energy and chemically stored it underground as their bodies were covered by dust, mud, and rock over the last 300 million years to form coal, oil, and natural gas. The byproducts introduced into the environment from converting these ancient stores of chemically sequestered solar energy into mechanical motion to power farms and factories, transport, and increase living standards are numerous. The ever-increasing release dosage and rate at which they are released into the environment raise concerns separate from those of running out of energy. Many researchers and scholars have that for the near and even midterms, there is no energy crisis, but if we are to minimize or altogether avoid future negative byproducts connected to past industrial practices, our relationship to abundant available energy must evolve.

Practical challenges – The energetic throughput and the associated inevitable losses of complex and highly ordered exergy with the simultaneous increase of low-order, random, entropy are important to monitor. Civilization thrives with abundant, well-ordered, high-quality energy sources available and becomes more difficult to maintain without it. But equally important is the *damage* produced in the meantime through their transformation. Environmental systems, both natural and augmented, are increasing, unable to meet the needs of a growing population. We measure ecological degradation by both what is added and what is lost. We can record the increasing parts per million or billion of harmful substances in water or air samples and the decreasing number of species on earth or lack of fresh drinking water in a population. Naturally occurring and manufactured products alike move through time and space according to the physical laws that govern all matter and energy. As they flow and interact with each other and with the environment, individual substances produce both positive and negative byproducts that can and must be responsibly monitored and assessed.

References

Atkins P, Paula J de (2010) Atkins' physical chemistry. Oxford University Press, Oxford

Bailyn M (1994) A survey of thermodynamics. American Institute of Physics, Woodbury

Blaustein R (2016) The great oxidation event: evolving understandings of how oxygenic life on earth began. Bioscience 66:189–195. https://doi.org/10.1093/biosci/biv193

Carnot S, Thurston RH, Carnot RH, Kelvin WT (1897) Reflections on the motive power of heat [microform]: and on machines fitted to develop that power. Wiley, New York, pp 37–126

Chadwick J (1932) The existence of a neutron. Proc R Soc London Ser A Contain Pap Math Phys Chara 136:692–708. https://doi.org/10.1098/rspa.1932.0112

Clausius R (1867) The mechanical theory of heat, with its Applications to the Steam-Engine and to the Physical Properties of Bodies, trans. John Tyndall. London, p 357

Donahue WH (1992) Johannes Kepler new astronomy. Cambridge University Press, Cambridge/New York

Falkowski PG, Isozaki Y (2008) The story of O2. Science 322:540–542. https://doi.org/10.1126/science.1162641

Fermi E (1934) Versuch einer Theorie der β-Strahlen. I. Z Phys 88:161–177. https://doi.org/10.1007/BF01351864

Feynman R (1963) The Feynman Lectures on Physics Vol. I Ch. 4: Conservation of Energy. http://www.feynmanlectures.caltech.edu/I_04.html. Accessed 7 Mar 2018

Galilei G, Favaro A (2020) Dialogue concerning the two chief world systems. Digireads.com Publishing

Ganesan V (2018) Thermodynamics: basic and applied. McGraw-Hill Education, New York, NY, p 131

General Accounting Office, U.S. Comptroller (1982) DOE funds new energy technologies without estimating potential yields. US Department of Energy, US Congress (GAO/IPE-82-1), 173p

Honerkamp J (2002) Statistical physics: an advanced approach with applications. Web-enhanced with problems and solutions. Springer Science & Business Media, Berlin, Heidelberg, p 298

Jevons WS (1865) The coal question: an enquiry concerning the progress of the nation, and the probable exhaustion of our coal-mines. Macmillan, London/Cambridge

Joule JP (1850) On the mechanical equivalent of heat. Philos Trans R Soc Lond 140:61–82

Lindbo DL, Kozlowski DA, Robinson C (2012) Know soil, know life. Soil Science Society of Agronomy, Madison, pp 70–81

Lindsay RB (ed) (1976a) On the mechanical energies of the solar system, by William Thompson (Lord Kelvin) Transactions of the Royal Society of Edinburgh, April, 1854, and Philosophical Magazine, December, 1854 (Mathematical and Physical Papers, Vol. II., Article LXVI). In: Applications of energy: nineteenth century, 1st edn. distributed by Halsted Press, Stroudsburg, Pa. New York, pp 179–194

Lindsay RB (ed) (1976b) Studies in energetics: II. Fundamentals of general energetics by Wilhelm Ostwald. In: Applications of energy: nineteenth century, 1st ed. Distributed by Halsted Press, Stroudsburg/New York, p 339

Maxwell JC (2010) A treatise on electricity and magnetism: volume 1, reprint edition. Cambridge University Press, Cambridge/New York

Newton I, Motte A (Trans) (2018) The mathematical principles of natural philosophy: the principia. CreateSpace Independent Publishing

Odum HT (1995) Environmental accounting: emergy and environmental decision making, 1st edn. Wiley, New York

Sahney S, Benton MJ (2008) Recovery from the most profound mass extinction of all time. Proc Biol Sci 275:759–765. https://doi.org/10.1098/rspb.2007.1370

Schwab K (2017) The fourth industrial revolution. Currency, New York

Senior AE, Nadanaciva S, Weber J (2002) The molecular mechanism of ATP synthesis by F1F0-ATP synthase. Biochim Biophys Acta Bioenergetics 1553:188–211. https://doi.org/10.1016/S0005-2728(02)00185-8

Siemens W (1883) British Association for the Advancement of Science, Report of the British Association for the Advancement of Science. John Murray, London

Smil V (2008) Energy in nature and society: general energetics of complex systems, 1st edn. The MIT Press, Cambridge, MA

Srinivasan R, Moe K (2015) The hierarchy of energy in architecture: emergy analysis, 1st edn. Routledge, London

Stevens JC, Choo KK (1998) Temperature sensitivity of the body surface over the life span. Somatosens Mot Res 15:13–28. https://doi.org/10.1080/08990229870925

Valko M, Rhodes CJ, Moncol J et al (2006) Free radicals, metals and antioxidants in oxidative stress-induced cancer. Chem Biol Interact 160:1–40

World Energy Statistics I Enerdata. https://yearbook.enerdata.net/. Accessed 27 Sep 2020

Chapter 3
Trash Can Living

Abstract The byproducts of consumption constantly flow between ecosphere and technosphere. LCA environmental impact categories—including global warming, eutrophication, acidification, and stratospheric and tropospheric ozone—form the basis to evaluate massive torrents of material and energy and their potential to degrade ecosystems.

Waste is defined not simply as unwanted or unusable material but more broadly as "material in the wrong place at the wrong time." *Earth* is the first of three "trash cans" that humankind has created as a result of its inability to deal with its waste. Illustrated examples establish the amount of land available to support us, the general magnitude of the solid waste problem, and the inherent disparity between the two.

Air, the earth's atmosphere, is the second "can" used to take up the lighter-than-air waste byproducts created by human activity. The evidence of an increase in greenhouse gasses correlates to the pronounced rise of fossil fuel-driven industrial production since the late 1700s through the present day. CO_2, tropospheric and stratospheric ozone problems, and methane are discussed.

Water comprises the third and final "trash can" survey, from solid waste fouling our waterways that ends up slowly churning in the great Pacific gyre to agricultural and industrial chemical elements and compounds that reduce the ability of the oceans, lakes, rivers, streams, and underground aquifers to support life.

Externalizing what is considered of no economic value transforms the world's litho-hydro-atmospheres into receptacles for waste.

3.1 Environmental Impact of Exporting the American Way of Life

Maybe the most important feature of the American Way of Life today is that it is no longer only American. Over the last hundred years, we have been extremely successful in exporting a particular idea and amount of what individuals and families around the world need to live and to thrive. According to current projections, as China and India, the two most populous nations, continue to adopt this standard of living, we will strain an already taxed resource transformation and distribution system.

© Springer Nature Switzerland AG 2021
J. Cays, *An Environmental Life Cycle Approach to Design*,
https://doi.org/10.1007/978-3-030-63802-3_3

To simply state the problem:
As the global population increases and living standards continue to rise:

More *people* pay.
More *money* to use.
More *energy* to change.
More *natural raw material* into.
More *stuff* and more *experiences* that produce.
More *waste*.

Working backward through the list, we see what it takes to support this way of life. The broad outline shows final effects first:

More *waste* from.
More *stuff* and more *experiences,* derived from.
More *natural raw material*, for which we need.
More *energy* and exchange.
More *money* from.
More *people*.

Each stage relates to each of the others and contributes, alone and together, to changes we see on the planet as the world's growing population constantly takes in and gives back more and more.

The solid forms of the first three components—waste, stuff, and raw material—are relatively easy to grasp because we can see and touch them. They have physical presence. We can put them in a pile, hold them in our hands, place them on a shelf, or sell them on eBay. The last three are more difficult to comprehend. *Energy* flow can be metered and its effects measured but, as discussed in the last chapter, cannot be directly seen or quantified. *Money* is merely a symbol of value and a convenient way to exchange goods and services. Yes, we can hold a dollar in our hand, but the paper on which it is printed has little inherent worth or utility until we exchange it for something we want.

Since the whole system has been created to support *people*, who are at the same time an integral part of the current production network, it is difficult to separate human efforts and activities, which are at work within the system, from the benefits we individually and collectively derive.

3.1.1   Waste

The complex interrelationships of each of the above components comprise the modern linear global economy. The end product of our traditional linear industrial economy is *waste*.[1] We think of it as inherently dirty, stinky, and useless. It is unwanted,

---

[1] The alternative to a linear industrial economy is a circular industrial economy first described by Walter Stahel as a system of loops in his 1976 report *The Potential for Substituting Manpower for*

deemed worthless, sometimes disgusting, or even dangerous. We want to *throw it a-w-a-y*—the farther, the better.

The problem with this view of waste, and our relation to it, is that there is no *away*. If we continue to reinforce these ideas in our own minds every time we think about the byproducts of our living, if we use language to describe our waste in only negative and fearful terms, it is almost impossible to see how *some* of what we discard could ever be a useful asset of tremendous worth. Another problematic outcome of trying to throw away our waste is that we are displacing huge amounts of raw material from places on the planet, where they are not harming anyone, altering them, and putting them in other places where they are causing problems for living creatures, including us.

The term "waste" is a context-dependent value judgment. The same physical material, as it moves through time and space, changes value as we use it. Carbon atoms, stored in fuel sources deep in the earth, are considered valuable resources to be extracted and set free to combine with oxygen during combustion. The instant they move from a necessary component of refined hydrocarbon fuel, used to do work, to an exhausted byproduct combining with oxygen in the air we breathe, their value as an asset changes to a liability cost in the form of too much atmospheric CO_2. The same is true for nearly every other natural resource we extract (take in), use, and discard (give back) to maintain our way of life in a linear global economy.

In order to externalize the costs of dealing with what is left of a thing that is deemed of having no value, we treat the air, land, and sea as *three garbage cans* with limitless capacity to absorb what we do not want. When we do this, we create problems. *Throwing away* requires that we collectively maintain the illusion that the earth is bigger than it is and that there are many fewer of us on the planet than there currently are. The truth is, though, that in a closed system, where technology and ecology comingle, waste byproducts remain extremely close by suspended and circulating in solid, liquid, and gaseous media. This new pollution directly and indirectly impacts biota that evolved for millions of years in environments in which it was absent.

3.2 Earth

Air and liquid water flow. They evade our direct perception as objects. Not so with the earth. It is the terra firma on which we stand, walk, ride, build, and live. Its solid surface literally supports all life. At the bottom of the deepest ocean, almost 7 miles down, you touch earth. Mount Everest and other peaks reach up more than 5 miles into the sky. Nearly all people on earth live comfortably between these extremes,

Energy to the Programme of Research and Actions on the Development of the Labour Market, DGV, Commission of the European Communities, Brussels.

within a vertical zone of less than 2 miles above sea level. We live where oxygen is abundant, where we can breathe easily.

While we travel below and above of that range for exploration, recreation, or specialized work like mining or scientific research, most activities take place relatively close to sea level. The amount of land available in this fairly narrow range for all human activity is large, but not infinite.

From the earliest cartographers to today's advanced satellite imagery and statistical surveys, we have made maps in order to measure and quantify territories and boundaries and the features they contain. We know that one-quarter of the nearly 60 million square miles of the earth's land surface is found more than 2 miles up and that a third more is desert. What's left is about 25 million square miles or 16 billion acres. That gives each of the nearly 8 billion people on earth (7.815B in the fall of 2020) less than 2 acres of habitable land, once we subtract public amenities such as airports, farms, rivers, roads, parks, recreational facilities, supermarkets, and landfills that we all share (Worldometer 2020).

3.2.1   Our Global Footprints

A person's global footprint is the sum of all the resources that go into providing the things and experiences that person has come to expect. Tallies of the amount of land it takes to support a single person living the "American Way of Life" come to over 20 acres per individual. If these calculations are correct, we will need ten times the amount of land we currently have to support the lifestyles more and more people around the world increasingly demand.

The situation is complicated further now that we are producing more waste that has to be put somewhere. The collective "garbage can" is already overflowing. We can't find places to put our trash anymore. We handle it, manage it, juggle it, and in the end just stick it on top of one of the piles we have already started. When we cannot pile it any higher, we look for new places to create more piles or holes to fill with it.

How Much Waste Do We Produce?
My young daughter, at just over 10 years old, has produced over 300 times her bodyweight in garbage; this is equivalent to the weight of a family of Asian elephants (Fig. 3.1). The volume it takes up would fill almost half our house from floor to ceiling. By the time she graduates from college, our family of three will have produced enough garbage to fill our entire house five times over. At nearly 50 tons, this solid waste will weigh the same as a humpback whale (Fig. 3.2). This all has to be put somewhere. Little by little, with each thing we throw "away" to be trucked off and dumped in a landfill, we are reducing the amount of land available to do the other things we want and need to do.

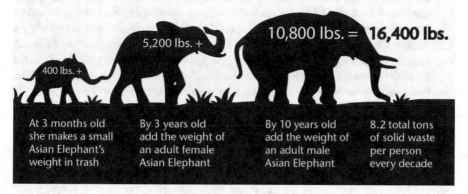

Three elephants = Total mass of garbage from one 10 year old girl

5,200 lbs. +

400 lbs. +

10,800 lbs. = 16,400 lbs.

| At 3 months old she makes a small Asian Elephant's weight in trash | By 3 years old add the weight of an adult female Asian Elephant | By 10 years old add the weight of an adult male Asian Elephant | 8.2 total tons of solid waste per person every decade |

Fig. 3.1 Individual solid waste graphic. (Data Source: US EPA)

Raising one child through college age, an American family of three produces over 50 tons of solid waste, equivalent to the mass of a humpback whale

Fig. 3.2 Family of three solid waste graphic. (Data Source: US EPA)

3.2.2 Pre- and Post-Consumer Solid Waste Components Travel

From the public's perspective, solid municipal waste is perhaps the most visible form of the problem. Consumers have direct experience with individual components handled at the moment they lose value. Solid waste byproducts at every previous stage supporting a product's existence up to that point can be orders of magnitudes larger and more harmful than the enormous waste footprint represented in the final disposal phase.

To accurately estimate the number of tons per person per year we must add to what goes into the landfills, the tonnage of all toxic or radioactive tailings produced and dumped into the landscape during the mining of raw materials, and the poison-ous industrial sludge and ash created in turning raw materials into finished goods. The total mass footprint of low or no value solid compounds extracted can be orders of magnitude greater than the item created. The environmentalist and documentary film maker, Chris Jordan, recently shared what he had learned on location in South America that "the mining footprint of a single gold ring is 40,000 pounds of dust, a pile of 40,000 pounds of toxic dust that is sitting on a hillside somewhere in Chile, and it's going to be leaching cadmium and mercury and arsenic into the soil for thousands of years" (Jordan 2020).

Through formal and accidental solid waste management practices, we are chang-ing the physical, chemical, and thermal profiles of ground we stand on. As the popu-lation grows, we will discard more and more. The lasting effects brought about by our increasing piles of detritus will continue to mount. Environmental impacts reach beyond local garbage dumps where, even under controlled conditions, seepage into groundwater and off-gassing of harmful compounds continue.

The systems currently in place in the United States to contain and treat the nearly 270 million tons of municipal solid waste we produce are complicated and expen-sive. We know that without them, some of the garbage won't just sit there. We line the bottoms of our dumps with plastic liners in order to keep "garbage juice" from seeping into groundwater supplies. We stick pipes into the layers of trash to capture methane from escaping into the atmosphere where, ton for ton, it is 30 times more potent than CO_2 as a heat trapping greenhouse gas. If this gas is not captured, treated, and used as fuel, it would join the earth's enormous natural stores of methane cur-rently escaping from previously frozen tundra soils. As global temperatures rise, these methane-rich ground sources, defrosting as fast as an unplugged freezer, will release a growing amount into the atmosphere, thus contributing to a vicious cycle of increasingly faster rising temperatures and even less permafrost to contain the ancient anaerobic decomposition byproduct.

3.3   Air

A clear majority of scientists regularly and unequivocally tell us that climate change is being driven by human activity. They identify increasing amounts of two primary greenhouse gasses, carbon dioxide and methane (Masson-Delmotte et al. 2018), that have been steadily added to the air for 200 years. Carbon dioxide, or CO_2, is produced when living beings breathe out. It gets produced at a much greater rate along with other long-lived greenhouse gasses when we burn things as we have for millennia and at a much faster rate during the last three centuries.

The US National Oceanic and Atmospheric Administration (NOAA) Annual Greenhouse Gas Index (AGGI) tracks the combined measurements of all long-lived greenhouse gases. In addition to NOAA's own global air sampling network in oper-ation since 1979, they use measurements of CO_2 going back to the 1950s from C.D. Keeling (1958), combined with atmospheric change evidence derived from air

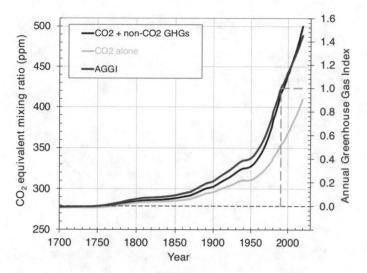

Fig. 3.3 Increases in greenhouse gas concentrations since 1950 have accounted for an overwhelming majority (72%) of the total increase in the Annual Greenhouse Gas Index over the past 260 years. (Graph courtesy of NOAA Earth System Research Laboratory (2019))

trapped in ice and snow above glaciers (Etheddge and Steele 1996). Equivalent CO_2 atmospheric amounts (in ppm) are derived with the relationship between CO_2 concentrations and radiative forcing from all long-lived greenhouse gases and shown in Fig. 3.3 (NOAA/ESRL 2020).

Most scientists agree that only in the last two centuries have CO_2 levels been rising outside of a "normal range." According to the evidence found in the historic record of ice core samples, the range in which human beings have lived and built our world civilizations since the middle ages has averaged just below 280 carbon dioxide molecules in every million molecules of atmospheric gas. We are now over 400 parts per million (PPM) and rising. According to NASA's former head of the Goddard Institute for Space Studies, Dr. James Hansen, we must reduce this level to 350 PPM in order to stabilize global temperature and the changes in climate that higher average temperatures bring (Hansen et al. 2008). This position has been peer reviewed and is supported by other scientists (Rockström et al. 2009).

3.3.1 Denial

Once we started extracting coal in large quantities and burning it to run factories over 200 years ago and, soon after, started distilling oil that would eventually be used to power everything else, CO_2 levels began to rise. As population increased in tandem with a growing industrialized economy in the last quarter of the nineteenth century, the rise in CO_2 became more pronounced. The following is one of the

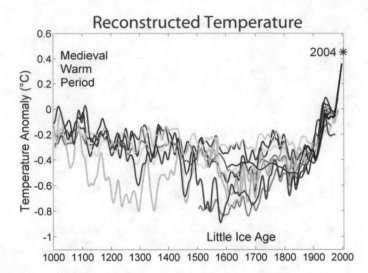

Fig. 3.4 Eleven different published reconstructions of mean temperature changes during the second millennium combined and published in 2005 by the physicist, Robert Rhodes. Used under Creative Commons Attribution Share-Alike License. Rhodes identifies 11 reconstructions used and differentiates them, in order from oldest to most recent publication using the following color key: (Dark blue 1000–1991): The Holocene. (Blue 1000–1980): *Geophysical Research Letters*. (Light blue 1000–1965): *Ambio*. Modified as published in *Science*. (Lightest blue 1402–1960): *J. Geophys. Res.* (Light green 831–1992): *Science*. (Yellow 200–1980): *Geophysical Research Letters*. https://doi.org/10.1029/2003GL017814. (Orange 200–1995): *Reviews of Geophysics*. https://doi.org/10.1029/2003RG000143. (Red-orange 1500–1980): *Geophys. Res Lett*. https://doi.org/10.1029/2004GL019781. (Red 1–1979): Nature. https://doi.org/10.1038/nature03265. (Dark red 1600–1990): *Science*. https://doi.org/10.1126/science.1107046. (Black 1856–2004): Instrumental data was jointly compiled by the Climatic Research Unit and the UK Meteorological Office Hadley Centre. Global Annual Average dataset TaveGL2v [1] was used

several "hockey stick" graphs that are now famous or infamous depending on which side of the Anthropocene argument you are on (Fig. 3.4).

Graphs like these are lightning rods in the debate between those who see what they show as proof of an existential threat to the human species and those who don't. Denialist arguments consider the graphs to be either false or, if true, inconsequential.

Starting with the 1998 Mann, Bradley, and Hughes graph, they may be the most politically controversial scientific graph ever produced, because they indicate correlations between human industrial production and a rise in global temperature. According to the graph, which reaches back to the Medieval Warm Period, the temperature did not vary above an average range before the early twentieth century, after the start of the Industrial Revolution (Mann et al. 1998).

Nearly everyone in the climate science community accepts these illustrations as indicators of how carbon-based industrial economic activities are directly causing climate change. They call for action to address root causes. Since acknowledging the scientific evidence of a problem would require a change in behavior to slow, stop, or reverse the flow of more CO_2 into the atmosphere, proponents of the growth of a carbon-based economy passionately argue that the graph itself is untrue. Critics of the graphs say there are serious problems with the underlying science and methods used to construct them and that there are records that actually show opposing (cooling) trends. Some say that maybe the graph is accurate but that atmospheric climate change is actually good for human beings for a host of either physical reasons or economic reasons. As criticisms are regularly addressed and more reconstructions corroborate the original findings (Masson-Delmotte et al. 2018), new ones spring up.

More important than what specific arguments are presented in the ongoing debate is to recognize the causal relationship between debate and inaction. A small amount of doubt introduced by a vocal minority makes reaching a consensus in clearly defining any problem to start with making it impossible to then move to address it. Doubt promotes a "paralysis through analysis" by extending the conversation indefinitely and delaying any large-scale action that would change the way we currently produce and consume. It creates the perfect excuse to not change anything until we know more; the thinking goes, "Until we are 100 percent sure, we should do nothing." (More on the denialist's views and tactics for sowing doubt for maximum effect in Chap. 6.)

3.4 China

As the global greenhouse gasses debate continues, the ill effects of business as usual are especially pronounced in rapidly industrializing countries. The Chinese government has acknowledged a growing problem. Its citizens are getting sick. The suffering caused by toxic smog is rising. As many as 3.7 million premature deaths in China were attributed to air pollution in 2012 up from 1.3 million deaths in 2008 (Mokoena et al. 2019).

Greenhouse gasses are not the only type of air pollution. The same industrial sources of carbon dioxide also produce levels of sulfur dioxide, nitrogen oxides, and ground-level ozone as well as other compounds in the form of fine dust that can collect in the lungs of anyone who breathes (Fig. 3.5). These unwanted ingredients in the air can trigger a host of maladies, according to Bai Chunxue, the head of respiratory medicine at Shanghai's Zhongshan Hospital. Diseases including lung cancer, chronic pulmonary obstructive disease (COPD), asthma, chronic bronchitis, and emphysema as well as cardiovascular disease are becoming more common in China (Demick 2013).

The same industries responsible for driving Chinese prosperity for the last three decades are causing one of the largest public health crises in the history of humankind. The government has pledged to improve the air quality in Chinese cities and major industrial centers but has not been able to make much progress. Even after technological advances developed to improve industrial emissions have been employed, the largest, centralized, government-run economy in the world cannot

Fig. 3.5 Visible evidence of suspended particulates and poor air quality in Eastern China. According to the photographer, both photos were taken in the "same location in Beijing in August 2005. The photograph on the left was taken after it had rained for 2 days. The right photograph shows smog covering Beijing in what would otherwise be a sunny day." (Photo credit Bobak Ha'Eri licensed under the Creative Commons Attribution-Share Alike 2.5 Generic License)

control rampant air pollution, which is largely a result of its growing need for electricity to power an ever-increasing number of factories and urban centers.

3.4.1   Off-Shoring: Far, but Not Away

Despite occasional trade war friction, China is the world's factory. It is a factory run on the country's most abundant energy source—coal. While China's economy grows, its people are asking why they cannot breathe clean air. Except for in the most densely populated industrial areas, most US citizens can take the air we breathe for granted. But, it wasn't always so. During the same period of time as China built its production capacity, the United States made the switch to a service economy and, along with many manufacturing jobs, effectively off-shored the majority of top polluting manufacturing industries. We sent many of our own pollution problems as far away as possible. We sent them to the other side of the world. Even though we have taken care of our immediate air pollution problems by relocating them, it has not been a complete success since climate issues caused by rising CO_2, NO_2, and other greenhouse gas levels are not confined locally and do not respect national borders. Another reminder that in a closed climate system, as easy as it is to think and act otherwise, there is no "away."

The spring of 2020 recorded a pronounced improvement in air quality across the globe. This coincided with a global pandemic that shuttered production facilities to, at once, protect workers from COVID-19 and respond to a cratering global demand in consumer goods. Satellite maps show the improvement from month to month in Eastern China (Fig. 3.6).

They also show year-to-year improvements between 2019 and 2020 Wuhan Province NO_2 levels. Pollution levels usually drop during the Chinese Lunar New

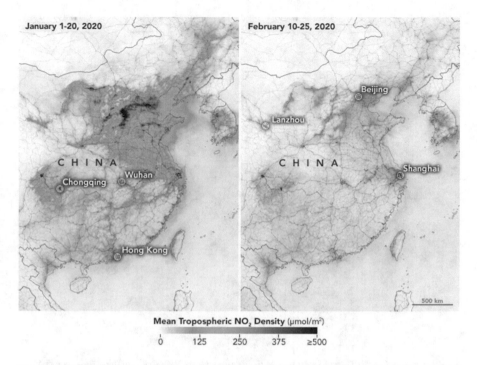

Fig. 3.6 NASA and European Space Agency (ESA) pollution monitoring satellites have detected significant decreases in nitrogen dioxide (NO_2) over China due to COVID-19 shutdown. (Photo credit: NASA)

Year Celebrations and then rise again in February when factories start production up again. In 2020 the levels did not rebound (Fig. 3.7).

3.4.2 Atmospheric Ozone

The atmosphere wraps the earth, and its currents are in constant motion, protecting all life from the vacuum of space and from multiple forms of radiation. It dampens the extreme temperature fluctuations between day and night. Temperatures on our own airless moon can swing more than 500 degrees Fahrenheit from a freezing night of minus 250 degrees to a boiling 250 degrees plus during the day.

The different layers of our atmosphere provide different kinds of utility and protection to the biosphere. Winds blow in three-dimensional currents reaching 10 miles above the earth's surface. Most weather activity occurs within the lower two-thirds of that region called the troposphere. Above that height is the next layer of the atmosphere called the stratosphere where the thin dry air is relatively still but stratospheric clouds occasionally form. For the next 10 or so miles up, oxygen gas molecules (O_2) are hit by ultraviolet radiation, breaking the bonds between the two

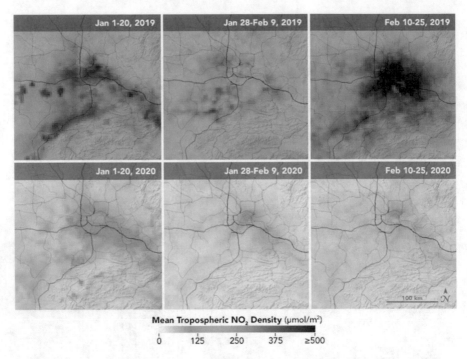

Fig. 3.7 Unlike in 2019, NO$_2$ pollution in 2020 drops in Wuhan and stays low after the Chinese New Year. NASA and European Space Agency (ESA) pollution monitoring satellites have detected significant decreases in nitrogen dioxide (NO$_2$) over China. (Photo credit: NASA)

atoms forming the oxygen molecule. The single freed atoms (O) then combine with unbroken O$_2$ molecules creating ozone (O$_3$) that make up the ozone layer. Ozone is unstable and rare. In every ten million air molecules, only an average of three are ozone molecules. Ninety percent of all ozone is found between 10 and 30 miles (16 and 50 km) above the earth's surface. This stratospheric ozone serves as a shield to the sun's high-frequency ultraviolet rays (Butler 2017). This ionizing radiation, when allowed to pass through the upper atmosphere unabated, can disrupt normal cellular functions causing skin cancer and other health problems in humans and many other organisms (NIEHS 2016).

3.4.3   Rational Global Policy and Action

In the 1970s and 1980s, reports of a growing hole in the ozone layer over the South Pole, and the crisis it represented, galvanized unanimous global action to ban the use of certain chemicals called chlorofluorocarbons that were used as refrigerants; as propellants in hairspray, deodorants, and other personal-care aerosol products; and as industrial blowing agents for foams and packing materials. In 1979,

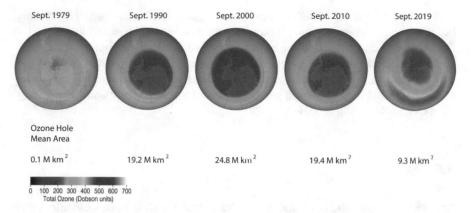

| Sept. 1979 | Sept. 1990 | Sept. 2000 | Sept. 2010 | Sept. 2019 |

Ozone Hole
Mean Area

0.1 M km^2 19.2 M km^2 24.8 M km^2 19.4 M km^2 9.3 M km^2

0   100   200   300   400   500   600   700
Total Ozone (Dobson units)

Fig. 3.8 The latest false-color view of total ozone over the Antarctic Pole compared to those over the previous views taken approximately 10 years apart shows a stabilizing of the ozone hole expansion. The purple and blue colors are lower ozone concentrations, and the yellows and reds show higher ozone concentrations. (Individual images courtesy of NASA, 2020)

predictions showed the ozone hole rapidly growing to 20 times its size, from 1.1 million square kilometers to 22.4 million square kilometers, by 1987. International governments listened to the science and took clear and definitive international action. To limit increased expansion of the ozone hole, they signed a binding agreement to limit the amount of ozone-depleting gasses each country could emit. Because of the relatively quick agreement and response by industry and government leaders with strong public support, the hole that was steadily growing 30 years ago has stopped expanding and is now expected to heal itself through the production of natural stratospheric ozone within the next 50 years. By 2000 the mean area of ozone hole was nearly 25 million square kilometers and by 2019 to 9.3 million square kilometers (Fig. 3.8) (NASA 2019).

If the arc of the ozone depletion crisis had followed the *current* trend in the larger climate debate, the rate of skin cancer and cataracts in humans would likely be much higher, as would reproductive abnormalities in numerous smaller animals and organisms. Living organisms from the largest to the smallest would have experienced harm. For instance, phytoplankton numbers would have dropped as an increased amount of harmful ultraviolet radiation penetrated into ocean waters. These simple single-cell organisms absorb as much as half of the CO_2 produced through burning fossil fuels and are a foundational component of the ocean food chain. As stratospheric ozone goes, so goes their and, eventually, our ability to live healthy lives. If the debate had continued past the 1987 Montreal Protocol signing, we might still be waiting for more evidence that an actual problem exists, considering what solutions would be best, or verifying that implementing the agreed upon solutions would not cause any negative economic effects. We would have been paralyzed by process and deliberation while being bathed in more dangerous UV waves with each passing day.

The ozone depletion crisis seems to be have peaked in the late twentieth century, but conditions require constant monitoring as new technologies are introduced to the market. The fact that the amount of protective ozone over Antarctica has measurably stabilized since just before the start of the millennium proves that our species has the ability to change climate destiny. We have not yet found the catalyzing argument to mobilize the world's decision-makers to effectively address the long-term CO_2 environmental crisis.

3.5   Water

The water we drink and cook and clean with is over 4.5 billion years old, older than our sun. It is not easy to destroy water. It takes a lot of energy to split one molecule of water into its constituent building blocks, that is, one oxygen atom bonded to two separate hydrogen atoms. Water endures through the three-phase changes we are all most familiar with and use each day—ice, liquid, and vapor. Nearly three-quarters of the earth's surface is covered with water, almost all of it salty. Over half of the typical human body is made up of water. Our dependency on clean fresh water to sustain life is clear. Without it, we would cease to exist.

In recent years, we have gained a deeper appreciation for just how much we depend on water to sustain our daily lives as we have failed to properly manage this vital resource. People live in places where water is growing scarcer year-by-year due to drought on the one hand and chemical contamination and biological waste pollution on the other. Moreover, we have used our waterways and oceans as a "convenient" place to discard things we no longer want.

3.5.1   Solid Waste Disposal into the Water

For much of the twentieth century, giant barges heaped with garbage were sent out to sea. They returned empty, ready to be filled again. While this practice is no longer routine in the United States and many other parts of the world, the flow of waste into the ocean has continued at an alarming rate. In addition, gravity and the flow of storm water through our streams and rivers, and to a greater degree through giant pipes and spillways, carry some refuse to our shores and out to sea. Even municipal storm and wastewater treatment systems, built to catch pieces of solid waste before they reach the ends of pipelines that lead directly to the ocean, miss a material percentage of what gets washed into and out of them.

Although we continue to build improved systems of collecting, treating, and filtering the waste we make, in the absence of funding for new infrastructure projects, we still must use leaky pipes built decades or even a century ago. As our systems age and the global population increases, our best planned and executed systems strain to keep pace. The bottom line is that we generate a greater total amount of waste than

Fig. 3.9 "The remains of dead baby albatrosses reveal the far-reaches of plastic pollution on Midway Atoll, 2000 miles from any mainland." Credit: Chris Jordan, from his series "Midway: Message from the Gyre." Used with permission under Creative Commons Attribution-Noncommercial-No Derivative Works 3.0 United States License. (Reproduced from chrisjordan. com, 2009)

ever before, and we have not been able to keep all of it from ending up in our streams, rivers, lakes, and oceans.

As a result, giant islands the size of Texas, discovered at the end of the twentieth century but accumulating long before, have formed in the Pacific Ocean. Wind and ocean currents allow these floating islands of garbage not only to persist but to continue to grow. As the churning flotsam breaks down over time, pieces of plastic, at various stages of weathering and disintegration, form three-dimensional columns of waste that extend down below the surface of the Pacific. There, fish and other marine life mistake the smaller particles for plankton and ingest them. Likewise, on the surface, hungry seabirds mistake colorful pieces of plastic for food for themselves and their young. Artist and film producer Chris Jordan's documentary, *Albatross*, focuses on a new generation of birds as they hatch and grow on Midway Island. He carefully traces the tragic distribution pathway of plastic flotsam that parents spend weeks at sea unwittingly gathering to feed their chicks (Fig. 3.9).

An estimated 225 million tons of plastic produced each year is contributing to the accumulating waste in both terrestrial and aquatic habitats worldwide. While the horrific effects of macroplastic debris on wildlife are well documented, the potential impacts of microplastic (<1 mm) debris which may account for 80% of all plastic stranded in the environment may have even longer lasting effects on the biosphere. Ingested microplastics can release toxic monomers that have shown to release toxic nonomers linked to cancer and reproductive abnormalities in humans, rodents, and invertebrates (Browne et al. 2007). As large a problem as plastic solid waste poses, it is only one of the many ongoing pollution crises affecting all life on earth.

3.5.2   Chemical Waste Runoff

Not limited to plastic bottles, bags, and wrappers that eventually float out to sea, our daily output includes other manufactured substances that contribute to dying waterways and oceans. The harm done by visible waste is easier to understand and react to than that caused less tangible flows. Our agriculture system is based on a foundation of chemical fertilizers that contaminate our waterways. The massive quantities of synthetic fertilizers used to replenish the nitrogen, potassium, and phosphates that crops draw from the soil and metabolize in order to grow are fossil fuel dependent. They require high-pressure, high-temperature, high-energy processes to produce and are used in greater amounts each year (Smil 2004). The monoculture cash crops of wheat, corn, soy, cotton, and softwood lumber all require massive quantities of fertilizer (and pesticides) annually to guarantee the harvest. Without steady chemical treatment of depleted soils, we cannot grow plants to feed and clothe and house ourselves. Some are spread as solid granules or sprayed as a liquid. In either form, they are applied to the surface of commercial cropland by the ton using trucks or planes. The soil, however, does not absorb these completely, nor do the crops use all of it. Strictly speaking, this fossil fuel-based way of feeding the plants that sustain us is inherently inefficient.

Excess synthetic fertilizer that is not taken up by the plants does not stay in place to be used the following growing season; it travels. Along with waterborne pollution from poorly managed sewerage and septic systems, some of it gets carried away when it rains either flowing downhill or, if the land is porous enough, through the hill. Flowing water, along with what it carries, also mixes and mingles downstream with more water. As rain carries leftover nitrogen, phosphorous, potassium, and other nutrients downhill, they can accumulate in the underground aquifers and contaminate wells from which we cook, clean, and drink. What excess that does not get pulled straight down into our groundwater in lower soil strata ends up as surface runoff in streams and rivers and eventually flows out into the bays, marshes, and other coastal waters.

3.5.2.1   The Effect of Nutrient Contamination in Water

The chemical elements that plants require to grow impact the well-being of other species as well. These elements are normally not found in abundance in waters where fish and other aquatic creatures live. When chemical nutrients, intended for surface plants, are introduced into these habitats, the effects are pronounced. Algae, unintentionally nourished by the excess nitrogen, potassium, and phosphorous in agricultural effluent, multiply in abundance. While alive, they can create a thick blanket on the surface that blocks sunlight that oxygen-producing underwater plants need to live. When the algae die, bacteria eat the remains. These bacteria reduce oxygen levels and produce carbon dioxide to a point where most aquatic animal life suffocates (NOAA 2020). Additionally, anaerobic bacteria that produce toxins can

Fig. 3.10 Eutrophication supports algae blooms in water. (Photo credit F. Lamiot, 2005 Attribution-Share Alike 2.5 Generic License)

flourish in this oxygen-depleted environment. Death spreads as the algae bloom (Fig. 3.10). This is called "eutrophication."

The global hydrospheric garbage can accepts whatever we intentionally or unintentionally attempt to throw away. Our waters, as vast and abundant as they are, are limited in what they can receive without negative consequences. Agricultural, industrial, and even domestic waste runoff ends up downstream in a poisonous soup incapable of supporting the range of life forms that naturally evolved under, atop, and beside the water's surface. We degrade our waters, normally teeming with life, through unchecked polluting. We impact an untold number of species, including our own, with steady streams of purposeful or accidental pollution.

3.6   Disordered and Harmful Out-of-Place Elements

There is nothing inherently bad or good outside a context. In terms of human survival and development, a thing becomes "waste" when it is not in a place that benefits humans. Any chemical, element, or compound, either naturally or synthetically derived, can be either good or bad depending on its context.

When a chemical element is a part of the soil, for instance, where a plant root can reach it, it acts as a necessary input for growing a plant that a human being will eventually use either as food or to make something out of. When an excess amount of the same single element escapes into a waterway, it can cause harm, perhaps destroying food sources like fish or rendering the water unfit for human consumption or even contact.

The plastic that ends up swirling around in the great Pacific garbage gyre served a purpose for someone at one time. Perhaps it was a plastic bottle filled with clean drinking water somebody needed for 10 minutes but then discarded. It was useful as a container until there was no water left in it. As soon as the person drinking from it was no longer thirsty, it became unnecessary. At that moment, it instantly became waste. The same is true for all food packaging as well as airborne and waterborne microplastic fibers used in clothing, cosmetics, and many other mass-produced consumer products (Cox et al. 2019).

Contrary to the perceptions of the average consumer, who casually discards it, the plastic bottle didn't disappear. It continued on long after it was empty. The same durability that held the water in without disintegrating keeps this synthetic compound from returning to its constituent parts, chemical elements available for use by other living beings without harming them. The plastic container might be buffeted by sun, wind, waves, and sand for years. It might get broken into smaller and smaller pieces of plastic that spread out for miles but remains, although fragmented and disbursed, essentially the same. Fish and birds that eat these plastic pieces are not fed by them. And these pieces and particles as well as other smaller foreign elements now exist where they were never found before. Living systems cannot integrate them; they can only do harm in inappropriate contexts.

3.6.1   Feedback Loops Affecting Solids, Liquids, and Gasses

Broadly dividing the earth's ecosphere into its discreet solid-lithospheric, liquid-hydrospheric, and gas-atmospheric components helps to conceptualize and measure the toll human activity takes on each of them. In every complex closed system, however, dynamic effects in one part of the system affect and influence those in other parts. As we produce and consume, the negative effects our waste creates in any one of the three "garbage cans" are constantly overflowing their boundaries into one or both of the other two.

Interacting positive feedback loops between the three have changed nature's self-regulating processes. The net effect of our actions, impacting the world's solids, liquids, and gasses that sustain all life, is what some label "anthropogenic" or human-caused climate change. Solid waste and liquid compound chemicals, which are not successfully sequestered, flow incessantly from our shores to pollute the oceans. They change the water's chemistry, kill aquatic life, and are adding a new stratum of human processed trash to the ocean floor. Some components of gaseous waste carried in the atmosphere, responsible for long-term climate change, migrate back into both solid ground and moving waters, causing other faster-acting problems through acidification.

Over months and years, the atmosphere mixes with the oceans. As moisture is continuously carried up through evaporation and returned by way of rainstorms on both land and sea, clouds become more acidic as they take up increased carbon dioxide, sulfur dioxide, and nitrogen oxides released from burning coal, oil, and natural gasses. The same substances that directly affect human health in the form of ground-level smog travel high into the atmosphere to mix with water and oxygen and other elements that eventually fall back to earth in the form of acid rain.

Rain mixes with soil and freshwater on land and salt water in the ocean increasing acidity over time. As soils become more acidic, their capacity to support plant growth diminishes. Harmful metals, stable in more alkaline soils, dissolve and are free to be taken up by plants as the pH levels in soil drop. As aluminum, mercury, and other heavy metals are released, trees weaken and forests suffer.

Oceans also act as one of the most effective sponges or "sinks" for airborne carbon dioxide. As oceans become more acidic, it makes it more difficult for marine organisms to form shells and skeletons that are primarily made of calcium, because calcium is dissolved by acids. These vulnerable species, which evolved in less acidic oceans, are now beginning to show severe signs of stress as alkaline habitats become less and less available. At some point, if the trends continue at the same rate, the acidity of the oceans will reach a point where calcium cannot coalesce to form shells or skeletons, and calcifying species, such as corals, clams, mussels, sea urchins, and others, will go extinct.

3.7 Conclusion

It is easy to think we can do something about the waste problems we face, that technological solutions are readily available, but effective tactics used to solve one problem may exacerbate another. Global climate summits set international targets for the net reduction of greenhouse gas emissions. Commitments are made. Some are kept and some are not. Some come with other side effects. In fact, one solution to reduce the use of common ozone-depleting chlorofluorocarbons increased the production and use of hydrofluorocarbons, which are powerful greenhouse gasses. Transitional technology strategies like these are stabilizing half-measures that often simply shift negative ecological impacts while we search for better, more encompassing solutions. Calcifying marine phytoplankton (Fig. 3.11), having dodged the UV bullet as

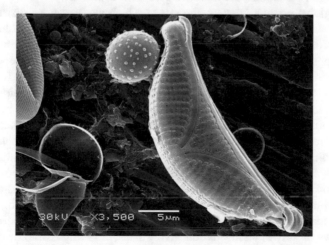

Fig. 3.11 The most diverse genus of phytoplankton is the diatoms, with an estimated 200,000 different species. Photo of diatom algae with spherical plant pollen. (Photo from Berezovska 2016, used under Creative Commons Attribution-Share Alike 4.0 International)

stratospheric ozone levels increase, may not be able to avoid the threat posed by a drop in the oceanic pH level.

When a species disappears, other living things are either directly or indirectly affected by its disappearance. Imagine that single species as a home to several different smaller organisms; as a predator or maybe a predator of a predator to another animal; or as prey for another that can no longer survive without it and whose own disappearance is felt higher up the food chain.

Whether collapse happens quickly or slowly by our standards, the end is the same, and it impacts the entire system. We cannot know the full effect of any given species as it interacts in pronounced and subtle ways in the food chain over time. We do know that a reduction of biodiversity makes the overall food system much less resilient. We do know that once a species is gone, it is gone forever. In the case of tiny phytoplankton, they form the foundation of the entire aquatic food web, *and* they process as much atmospheric carbon through photosynthesis as all the earth's forests, jungles, and other land vegetation (Armbrust 2009). Even those who do not eat seafood must still breathe air.

We are changing the biosphere's equilibrium. Our actions impact the fundamental conditions supporting all life on earth. The physical and biological systems, which have been continuously developing over millennia and on which we have become so dependent for our own survival, are changing in complex, subtle, and deep ways; we cannot foresee all the outcomes.

As we extract and transform natural materials, we inevitably increase what goes to the three garbage cans. Because materials flow, the things we throw away eventually intermingle with the natural environment. Elements, compounds, and the products we make out of them flow through the air and through the water, and over time, they penetrate and even form parts of the solid earth beneath us. The more materials we take in and use to support the consumptive habits we have developed and passed down through generations, the more we give back as waste.

As the global populations rise, the rate at which we have been distributing our increasing waste into the "three garbage cans" also rise. The sheer volume of it is producing chemical and physical changes in the environment registering at a scale that no other single species can match. The magnitude of changes that coincide with the inflection point in "hockey stick" graphs defines a distinct period in geologic time—the last 200 years—the Anthropocene, in which human activity has influenced climate and environment.

The record of our global production and consumption activities that define the Anthropocene is being meticulously captured by the earth's gravity. Each of the trillions of particles, thrown up into the air or flowing into the water, eventually settles on the earth's surface. There is no way of erasing the physical evidence of our lives and actions. Deposits of artificial compounds and increased carbon, radioactive isotopes, and plastic waste will be sequestered for eons in a thin layer of the earth's crust to be read as distinct from the rest of the Holocene interglacial period.

While we are alive, taking in and giving back, it is impossible to do nothing. Even without ascribing moral labels of "good" or "bad," our actions have consequences. When we all tacitly agree to do nothing to reduce the amount of these compounds that enter our air and our waters, we now do it with the full knowledge that the effects are not neutral nor are they simple.

The next chapters will present methods to first understand what we can measure and track and offer approaches to act, to make informed decisions.

References

Armbrust EV (2009) The life of diatoms in the world's oceans. Nature 459:185–192. https://doi.org/10.1038/nature08057

Browne MA, Galloway T, Thompson R (2007) Microplastic—an emerging contaminant of potential concern? Integr Environ Assess Manag 3:559–561. https://doi.org/10.1002/ieam.5630030412

Butler JH, Montzka SA (2017) NOAA/ESRL global monitoring division - THE NOAA Annual Greenhouse Gas Index (AGGI). In: THE NOAA Annual Greenhouse Gas Index (AGGI). https://www.esrl.noaa.gov/gmd/aggi/aggi.html. Accessed 14 Dec 2017

Cox KD, Covernton GA, Davies HL et al (2019) Human consumption of microplastics. Environ Sci Technol 53:7068–7074. https://doi.org/10.1021/acs.est.9b01517

Demick B (2013) Lung cancer: a cloud on China's polluted horizon. In: Los Angeles Times. https://www.latimes.com/world/la-xpm-2013-dec-24-la-fg-china-lung-cancer-20131224-story.html. Accessed 8 Jul 2020

Etheddge DM, Steele LP (1996) Natural and anthropogenic changes in atmospheric CO2 over. J Geophys Res 101:4115–4128

Hansen J, Sato M, Kharecha P et al (2008) Target atmospheric CO2: where should humanity aim? Open Atmosp Sci J 2:217–231

Jordan (2020) Can beauty save our planet? | Chris Jordan | TEDxSeattle. Seatle. WA

Keeling CD (1958) The concentration and isotopic abundances of atmospheric carbon dioxide in rural areas. Geochim Cosmochim Acta 13:322–334

Mann ME, Bradley RS, Hughes MK (1998) Global-scale temperature patterns and climate forcing over the past six centuries. Nature 392:779–787. https://doi.org/10.1038/33859

Masson-Delmotte V, P. Zhai, Pörtner H, et al (2018) Summary for Policymakers. In: Global Warming of 1.5°C. An IPCC Special Report on the impacts of global warming of 1.5°C above pre-industrial levels and related global greenhouse gas emission pathways, in the context of strengthening the global response to the threat of climate change, sustainable development, and efforts to eradicate poverty. https://www.ipcc.ch/sr15/chapter/summary-for-policy-makers/

Mokoena KK, Ethan CJ, Yu Y et al (2019) Ambient air pollution and respiratory mortality in Xi'an, China: a time-series analysis. Respir Res 20:139. https://doi.org/10.1186/s12931-019-1117-8

NASA (2019) 2019 Ozone hole is the smallest on record. https://earthobservatory.nasa.gov/images/145747/2019-ozone-hole-is-the-smallest-on-record. Accessed 27 Sep 2020

NIEHS (2016) National toxicology program: 14th report on carcinogens. In: National Toxicology Program (NTP). https://ntp.niehs.nih.gov/go/roc14. Accessed 5 Dec 2020

Rockström J, Steffen W, Noone K et al (2009) A safe operating space for humanity. Nature 461:472–475. https://doi.org/10.1038/461472a

Smil V (2004) Enriching the earth: Fritz Haber, Carl Bosch, and the transformation of world food production. The MIT Press, Cambridge, MA/London, pp 61–82

US Department of Commerce N, Butler JH, Montzka SA NOAA/ESRL Global Monitoring Laboratory – THE NOAA ANNUAL GREENHOUSE GAS INDEX (AGGI). https://www.esrl.noaa.gov/gmd/aggi/aggi.html. Accessed 11 Jul 2020

US Department of Commerce NO and AA What is eutrophication? https://oceanservice.noaa.gov/facts/eutrophication.html. Accessed 27 Sep 2020

Worldometer (2020) World population clock: 7.8 billion people (2020) - Worldometer. https://www.worldometers.info/world-population/. Accessed 27 Sep 2020

Chapter 4
Do Something: Mid-twentieth Century Developments

Abstract Life cycle thinking as it pertains to cost through time is developed for the US military industrial complex. An analyst working in the late 1950s for the military contractor, RAND, introduces and applies the life cycle concept and term to nonliving things. By the late 1960s and early 1970s, popular sentiment, science, and government (EPA) use it as a structure in the modern environmental movement.

Increased awareness of solid waste's many negative and invisible health and environmental impacts and the problem of its persistent physical presence in the landscape spurred people to action. Less visible than solid waste but equally critical, air- and water-quality problems require increased government regulation to monitor pollution and create policy to hold industry accountable. By the end of the decade, corporations began to formally self-assess their role in the problem and tie improvements in environmental performance to operational efficiency and build the precursors to today's LCA models.

Homo faber as environmentalist emerges in the twentieth century able to take action to effect limited positive change in large-scale resource management. Post-World War II responses to a growing solid waste crisis were blunted throughout the twentieth century by institutionalized programs of planned obsolescence and a growing consumer class.

4.1 Purpose-Driven Managed Resources

Most remarkable stories of benefits realized through *doing something* have a clear goal driving the action. One example comes from popular reduction, reuse, and recycling of resources to support wars backed by overwhelming popular support. Some people still remember the practice of stockpiling materials that were in high demand and short supply during World War I and World War II. From carefully keeping and turning in everything from peach pits to be processed into carbon filter material for gas masks to tin cans to be used to manufacture weapons for the fighting troops throughout the first half of the twentieth century, the practice was tied to patriotism and a collective goal to win two World Wars. A clear and present danger in the form of a foreign enemy gave purpose to the immediate task at hand and engendered focus and discipline on a mass scale. It's an example of how a compelling common purpose can coalesce millions of separate actions into a collective

J. Cays, *An Environmental Life Cycle Approach to Design*,
https://doi.org/10.1007/978-3-030-63802-3_4

mission. Rationed goods during World War II created an entirely new scrap metal and textiles collection and reselling industry (Woods and Petersen 1995).

During the first half of the 1940s, most people responsible for managing and contributing to this effort had firsthand experience with personally managing scarce resources. Most had lived through the great depression and were already personally inclined to save every scrap of material they had. These habits were born out of a sense of individual self-preservation from living through years of deprivation.

It was not difficult to cultivate appreciation in the citizenry as a whole for what each and every individual piece of metal, fabric, paper, glass, or rubber could mean for the day-to-day success or failure when so many knew personally what it was to live without enough material resources. The concerted effort lasted for a few years, especially toward the end of the war as demand for diminishing resources increased, and ended with a decisive victory and a sense that it was all worth the effort.

Once the war was won, however, it was no longer necessary to continue to scrimp and save nor for everyone to focus on individual fragments and leftovers. Since the troops were home, the people of the American public could turn its focus from the manufacture of implements of defense and destruction to other, more productive, endeavors. And so, as a nation, they did.

4.2   New Priority: Consumption

The years following the end of the war ushered in a period of mass consumption in the United States and around the world unparalleled in human history. A new sense of stability, security, and prosperity inspired optimism, and new appetites developed for more than a subsistence level of existence. Positive popular sentiment inspired by a sense of security and perceived abundance drove the sharp rise in population known as the "baby boom."

Growing families increased demand for just about every conceivable good and service. The World War II military-industrial complex, set up to win the wars abroad, had the enormous capacity needed to continually supply the troops. As they returned home to have families and enjoy the fruits of a hard won peace, these industries retooled. Companies that made tanks, fighter planes, and gunboats retooled to produce cars, passenger planes, and powerboats for recreational use to satisfy the demands of a growing and more affluent consumer and leisure class. The waste problems created by this post-war euphoria consumption would need a systems-level approach to understand and quantify.

A recounting of a few key events in the next section traces the emergence of the "life cycle" concept that today is an easy take as a matter of course in describing any phenomena that change through time.

4.3   Military Roots of the Modern "Life Cycle"

"Life cycle" naturally captures the changing qualities of living beings throughout their existence. Referring to a nonliving object's or system's life cycle moves beyond the poetic devices of personification or anthropomorphization. These literary techniques imply living behaviors in things that are not alive but rarely are mistaken for a description of something actually coming to life in the biological sense. While similar to Hobbs' use of the Leviathan and the Behemoth to condense sprawling and complex political structures made up of living people and living laws into images of gigantic biblical monsters, the term, borrowed from the biological sciences, avoids the spectacular or the fantastic. The reference in this modest budgeting report to an analytical process description, instead, transfers to nonliving material some of the essential qualities of living beings.

A cynical interpretation concludes that, with this conceptual transference, the animate and inanimate are inadvertently conflated or equalized in terms of how they can be accounted for and ultimately valued. The uncomfortable question of the monetary value of human life is, however, already answered in the actuarial tables kept by insurance companies, and these tables date from the seventeenth and eighteenth centuries.

A more optimistic view of the same developments could anticipate that, in the coming decades, we might be able to interact with objects and systems throughout their entire life spans to support and improve the quality of life and that, as cybernetic elements, each produced thing would be able to provide feedback regarding their status at key moments. Sensing and communicating, acting in and reacting in relation to their context, they could provide metrics and insight about a variety of critical aspects regarding their relationship to us and our shared environment over time. This would partially manifest early systems dynamics implied ideals of building a network of self-regulating components all working individually and together for the good of humankind.

The actual term "life cycle" has been in use since the early nineteenth century in Germany to describe the natural rhythms of biological systems from birth to death. Science writers often mixed descriptions of life and its changing phases with references to the divine:

> The Creator caused the infinite number of varied, beautifully formed, flowering, and fragrant plants to sprout from the ground, not to blossom and wither, but to be useful in many ways to man and to the whole organic kingdom; he gave the mind to use them expediently, to observe them in the various periods of their metamorphoses, before and after their *cycle of life*, in order to increase their knowledge, to cultivate their mind and to fulfill its destiny.
>   John, JF (1810). Chemical investigations of mineral, vegetable and animal substances. Continuation of the chemical laboratory, vol. 1: 5. (Toepfer 2018)

In the late1950s, a century and half after what may be the earliest published reference of the term used in biological sciences, the life cycle concept would be appropriated and officially applied to nonliving things and, perhaps most ironically, to the financial analysis of national weapon systems and armaments. This proposition would be quickly accepted and fully codified by the US Government's industrial procurement complex and normalize it for broader use later on.

The US Department of Defense (DoD) is the largest and oldest government organization in the country. Evolving out of the early observed practical need for effective armed forces to defend the United States even before they were formally constituted, the DoD is now one of the world's largest single landholders, with approximately 28 million acres under its control in the United States, its territories, and in nearly 40 countries (Brady 2014). In the first decades of the twenty-first century, it oversees well over half a trillion dollars annually and is responsible for purchasing everything needed to provision all branches of the US military and managing the physical resources to support national defense.

In 1798, the US Congress created the War Department and Navy Department, which would operate as independent agencies until they were merged due to national budgetary realignments the mid-twentieth century that created the current US military industrial complex. Just after World War II, President Harry Truman, in a December 19, 1945, speech to Congress, formally identified the need to reduce duplicative and wasteful military spending of the nation's limited financial, physical, and human resources caused by interdepartmental conflicts. In his *Special Message to the Congress Recommending the Establishment of the Department of National Defense*, Truman set a national priority to "realize the economies that can be achieved through unified control of supply and service functions" (Truman 1945).

Adding support to this presidential announcement regarding budget consolidation, then Army Chief of Staff, General Dwight Eisenhower, proposed establishing formal ties between military and civilian agencies. Academic and industrial partners would be supported to spur creativity and innovation in broad areas of scientific and technological inquiry to "directly benefit the Army" and "contribute to the nation's security by cultivating 'friendships invaluable for future cooperation'." In order to start outsourcing advanced technological development, Eisenhower's April 27, 1946, memo to the Directors and Chiefs of the War Department and others said:

> The Army must have civilian assistance in military planning as well as for the production of weapons Effective long-range military planning can be done only in the light of predicted developments in science and technology. As further scientific achievements accelerate the tempo and expand the area of our operations, this interrelationship will become of even greater importance. In the past we have often deprived ourselves of vital help by limiting our use of scientific and technological resources to contracts for equipment. More often than not we can find much of the talent we need for comprehensive planning in industry or universities. Proper employment of this talent requires that the civilian agency shall have the benefit of our estimates of future military problems and shall work closely with Plans and the Research and Development authorities. A most effective procedure is the letting of contracts for aid in planning. The use of such a procedure will greatly enhance the validity of our planning as well as ensure sounder strategic equipment programs. (Marcy 1980)

Two years after his first public recommendation, Truman signed the National Security Act of 1947. This act established the formation of the Department of Defense (DoD) as well as the Central Intelligence Agency (CIA) and the National Security Council (NSC) in order to not only increase efficiency but also efficacy of the national security apparatus in its primary mission to defend the US Constitution and its physical territory. The single entity now also had a larger and more influential profile.

4.3.1   RAND and General Systems Theory

Coinciding with Truman and Eisenhower's consolidation of military branches and closer integration with outside private research and development entities, between 1945 and 1948, the RAND Corporation was established. RAND, whose name is a contraction of "Research and Development," began as a small, specialized US Army Air Force contract project within the Douglas Aircraft Company in order to increase the application of advanced scientific research to improve national security. In just 3 years, RAND incorporated in Santa Monica, California, as an autonomous non-profit Think Tank with 200 employees working collaboratively on dynamic multi-variable systems problems at the nexus of mathematics, engineering, aerodynamics, physics, chemistry, economics, and psychology. During this early period, transdis-ciplinary teams in mathematics and mathematical economics advanced the fields of modeling and simulation including Monte Carlo methods, linear and dynamic pro-gramming, game theory, network theory, and cost analysis to build a "rational" method to quantify and evaluate the expected costs, benefits, and risks through a comprehensive systems analysis approach built to "address military decision prob-lems mathematically" (Jardini 1996).

Systems analysis is a subset and an application of general systems theory (GST). The Austrian biologist, Ludwig von Bertalanffy, originated and developed his "Allgemeine Systemlehre" in the interwar period and brought it to the United States during the War. Bertalanffy's ideas and work would also directly influence ecological systems engineer and scientist Howard T. Odum who would then directly influence the earliest industrial adopter of proto life cycle assessment, discussed later in this chapter.

As a small detail in its applied systems analysis approach to solve complex prob-lems but keeping with the spirit of Truman's original decision to merge all military branches to improve "supply ad service functions," RAND created and socialized an expanded and durable "life cycle" definition, quite removed from any original references to biological organisms and networks. This implied extension to apply the meaning to nonliving things lies at the heart of all modern life cycle assessment activities.

A decade after RAND began operating as an independent DoD consultant, David Novick, working in RAND's Cost Analysis Department, appropriated the biological concept of "life cycle" and applied it to inanimate defense systems (Fig. 4.1). Common government budgeting practice at the time counted only immediate costs expected over one "future fiscal year." Novick argued for the use of an expanded time scale in order to accurately assess the true total financial costs associated with a specific weapon system. He also introduced the notion of discreet phases and broke down typical funding patterns over "a long span of years" (Novick 1959).

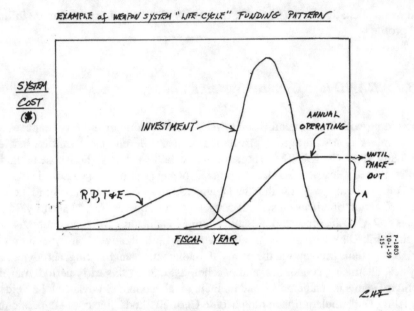

Fig. 4.1 Hand illustration included in David Novick's RAND report illustrating the three main phases of weapon system life cycle funding pattern. The three phases are (1) research, development, test, and evaluation (R,D,T&E), (2) investment, and (3) annual operating. The third phase continues until the asset's end of life. (Used with permission from the RAND Corporation)

This pragmatic, conceptual, life cycle costing model included many of the realistic caveats and challenges with which later life cycle assessment models would have to contend. These included the model's inherent uncertainty levels, potential lack of precision, and inability to accurately predict every event that would affect cost over a multi-year time span. Therefore, rather than attempting to foresee all details and contingencies and then commit to a fixed budget according to a single scenario, Novick recommended that government funding appropriations continue year by year.

The unstated assumption was that there would be incremental oversight by responsible authorities to make funding changes in line with these weapons systems annual operational requirements. He was essentially recommending a flexible and iterative process that would create a more inclusive and therefore accurate estimation of expenses. This early picture of total costs throughout the useful service period of a proposed system would better inform the decision-makers entrusted with the responsibility of wisely spending tax dollars.

4.3.2   End of Life and Early Obsolescence

So-called boneyards were massive sites where out-of-service war time material was collected, inventoried, stored, stripped down, and recycled. This would be where all the budgeted equipment would go in the last phase after the "phase out" was complete. The informal but popular moniker given to these facilities subtly elevates the sites from mere scrap facilities to the final resting places for the remains of once vital beings. The name casually given to these sites in the mid-1940s foreshadows Novick's metaphorical transference that, a decade later, would blur the distinction between living beings belonging squarely to the ecosphere with the inanimate ones of the technosphere.

Stockpiles of leftover materials were no longer needed to feed the fight. Instead, they were used, for pennies on the dollar, to jumpstart a new consumer economy. Besides the general-purpose vehicle (GP) or jeep, few products successfully made the direct transition from the war theater to the consumer market, so a lot of scrap metal had to go somewhere or sit and rust. Steel and aluminum, two of the most malleable and easily recycled materials available, can be melted down and reformed, repeatedly, without losing any of the performative properties that make them useful. The Kingman Army Airfield in Arizona alone became the site for scrapping and smelting 57 million pounds of aluminum and 21 million pounds of steel from airplanes over an 18-month contract with the Wunderlich Company in 1945 and 1946 (airplaneboneyards.com 2020) (Fig. 4.2).

Fig. 4.2 Recycling obsolete equipment at the Kingman Army Airfield in Arizona, commonly known as the "boneyard." (1947 Photo by Peter Stackpole. Licensed for editorial use by Getty Images from the LIFE Picture Collection)

Thus, furnaces continued to work day and night to turn war machines into commercial, industrial, or domestic goods. Swords were turned into ploughshares and tanks into toasters. The redeployment of recovered metal atoms was not driven primarily by a sense of stewardship of resources or an environmentally sustainable ideal. It was the market that recognized the potential in obsolete war machines as the ready feedstock of a new economy that drove recycling materials at this time.

A thing becomes obsolete when it can no longer serve the purpose for which it is made. War, by its very destructive nature, increases the rapid obsolescence of everything produced to support it. The largest war production machine ever built, now converted to produce in a time of peace, had to change both what it made and the way it made it. The terms "durable goods" or "hardware" have different meanings to someone in battle than to a consumer out shopping for kitchen appliances. In the absence of constant destruction on the battlefield, obsolescence became a feature that had to be designed into every new product for sale in the post-war economy. Dynamic obsolescence could be designed through the eventual mechanical failure of a key product component that fails through normal use before the rest of a product does. More commonly, a consumer product is obsolete not when it physically fails but when cultural forces dictate it is no longer suitable—when it goes out of fashion.

In order to keep companies in business making things, more people had to do something the opposite of what they had just finished doing. They had to buy more of what the factories produced. And in the quarter century following the War, Americans did just that. The popular pre-war attitude, governed by a sense of scarcity, gave way to one that celebrated the abundance available for the modern consumer class.[1]

The notion of duty to country remained. It was also converted from a war time virtue to one that fueled consumption. However, to do one's duty as an American now meant to, as my grandmother would regularly lament, "Throw out and buy new!" American consumers, many of the same people who won the war through sacrificing, going without, and giving nearly everything they had, needed to be convinced to adopt new attitudes and new habits. Whether explicitly or implicitly, American post-war industry and the consumers buying from it were in tacit agreement to meet a certain amount of demand from the supply that was readily available and to not let it go to waste.

---

[1] The concept of planned or dynamic obsolescence was introduced in the 1920s and 1930s but was popularized by the industrial designer, Brooks Stevens, in 1954.

4.4   Corporate Origins of Environmental LCA

In order to convince those belonging to "the greatest generation" to behave differently and bring up their children as consumers, who would support the economy and spur new growth each year, new media outlets spread a message designed to entice, to seduce, and to build a sense of entitlement and want. Twenty-five years after the end of World War II, according to this new code of conduct, Americans were largely dissuaded from anything we would recognize now as the practice of reducing, reusing, or recycling on an industrial scale.

The cultural turmoil of the 1960s introduced deep questioning of established practices and societal norms especially around consumption that bolstered post-war consumerism. Thinkers and social activists articulated alternative behavioral principles of the US countercultural revolution and spurred people to act. The ensuing protests reached a crescendo during the Vietnam War. Greed and the violence it engendered was blamed on the establishment. The overconsumption and resulting increase in pollution prepared the way for the modern environmental movement.

By 1970, the environmental crisis was evident to a much larger proportion of the consuming population. The US mass media began to reflect new trends. Pollution that was easily seen, smelled, and felt in the smog that blanketed most cities and industrial towns, rivers that caught fire due to floating industrial sludge, and massive, open garbage dumps close to many rural and urban communities around the country were the constant target of popular criticism. Once people made the direct connection back to commerce and industrial output, elected political leaders as well as corporations could not ignore the public outcry. They had to act in response to growing pressure to address the problems of a deepening environmental crisis.

4.4.1   The Coca-Cola REPA

Ten years after the introduction of the "life cycle" concept into the federal budgeting process for military weapon systems procurement and maintenance, the world's largest soft drink corporation developed the conceptual prototype for environmental life cycle modeling. The approach does not seem to be directly inspired from governmental life cycle budgeting practices but does share its origins in general systems theory.

In 1961, industrial engineer Harry E. Teasley, Jr. joined the Coca-Cola Company as a senior engineer in the Technical Research and Development Department. By 1969 he was managing Coca-Cola's packaging functions[12] (Plastics Institute of America 1991). Teasley's innovative contribution grew out of a desire to understand how Coca-Cola's containers broadly impacted the environment. Beyond

---

[2] By 1987 Teasley would lead Coca-Cola Foods, a division of the Coca-Cola Company, as President and CEO.

considering the impacts related solely to the visible solid waste disposal problem, the major crisis of the day in 1969, he envisioned a study that would simultaneously assess multiple impact categories.

Coca-Cola may have lacked the in-house expertise and staffing to carry out to fruition what had not been done before, but they possessed the financial resources to fund a groundbreaking study. The research team would have to adequately address inherent complexity involved in defining, collecting, and analyzing the data on multiple interrelated processes. In the spring of 1970, Teasley contracted Midwest Research Institute (MRI) to conduct the first formal commercial "resource and environmental profile analysis" (REPA). The REPA established the basic framework for what would, 20 years later, become life cycle assessment (LCA). MRI's Assistant Director of the Economics and Management Science Division, Arsen Darnay, collaborated with Teasley to design and execute the study to meet two objectives:

1. To establish the total environmental impacts created by the use of each of several selected container systems available to Coca-Cola USA; and
2. To compare the impacts of various types of containers under the assumption that each container type is used exclusively for Coca-Cola packaging. (OTA 1976)

In the final articulation these *impacts* would include:

1. Materials consumption
2. Energy consumption
3. Water consumption
4. Solid waste generation
5. Energy effluents
6. Air pollutant emissions
7. Waterborne waste emissions
8. Economic impacts

Eight container *system* types were compared:

1. Electrolytic tin plated (ETP) can
2. Tin free steel (TFS) can
3. Aluminum can
4. One-way glass bottle (OWB)
5. Returnable bottle (RB)
6. Owens-Illinois glass composite package (GCP) container
7. Polyvinyl chloride (PVC) bottle
8. Monsanto "Lopac" bottle

They would be assessed over five *operations* phases:

1. Mining
2. Processing
3. Fabrication
4. Filling
5. Consumption/discard

4.4.2 Two REPAs Measure Similar but Not the Same Indicators

Robert Hunt and William Franklin, writing in the first issue of the International Journal of Life Cycle Assessment, recount how Coca-Cola's Teasley and MRI's Darnay originally set the foundations of LCA. Hunt and Franklin focus on the pre-1990's history of LCA as it quickly moved from a commercially driven endeavor to a standard global analytical method. MRI would conduct many more REPAs for Coca-Cola and other private companies throughout the next two decades. Most industry-contracted REPA study results continue to be kept private. This is true for both publicly or privately held company studies, which are likewise based on detailed process and raw material descriptions that could be used by industry competitors.

A few years after the original proprietary study was conducted, Darnay, who had moved on to serve as Deputy Assistant Administrator for Solid Waste Management in the early years of the EPA, worked again with MRI, this time on a publishable study of beverage container alternatives that shared the technical details not available in the original commercial study. Hunt and Franklin, while not named investigators in the first study, were working at MRI at the time and were later named principal investigators conducting the US Government's *Resource and Environmental Profile Analysis of Nine Beverage Container Alternatives*.

Their final 1974 REPA report evaluates a slightly different collection of containers (for beer this time) than the initial Coca-Cola study:

1. Glass bottle—returnable 19-trip—used on premises
2. Glass bottle—returnable 10-trip—refilled at a central bottling facility
3. Glass bottle—returnable 5-trip—refilled at a central bottling facility
4. Glass bottle—one-way
5. Glass bottle—plastic coated
6. Steel can—three piece with aluminum closure composite
7. Steel can—all steel
8. Aluminum can—two piece all aluminum (15% of cans recycled)
9. Plastic bottle—ABS plastic

Included in the study was a section that analyzed environmental impacts for soft drink containers as well. This list was also modified from the original study and included:

1. Glass bottle—returnable 15-trip
2. Glass bottle—one way
3. Glass bottle—plastic coated
4. Steel can—three piece with aluminum closure composite
5. Steel can—all steel
6. Aluminum can—two piece all aluminum (15% of cans recycled)
7. Plastic bottle—ABS

Side-by-side comparisons of the Coca-Cola and EPA studies show that impact categories in the EPA study were modified and, most importantly, that the scope of the government-funded study omitted a key category that was included in the corporation-funded study:

MRI Coca-Cola study	MRI EPA study
1. Materials consumption	1. Raw materials—kg
2. Energy consumption	2. Energy consumption—10^9 joule
3. Water consumption	3. Water consumption—10^3 liter
4. Solid waste generation	4. Post-consumer solid wastes—cum
5. Energy effluents	5. Industrial solid waste—cum
6. Air pollutant emissions	6. Atmospheric emissions—kg
7. Waterborne waste emission	7. Waterborne wastes—cum
8. Economic impacts	

The 1974 EPA study largely follows the original Coca-Cola study's definitions for major environmental impact categories: material, energy, and water consumption and solid, waterborne, and atmospheric waste generation, but dispenses with the eighth category, "economic impacts." In a purely environmental profile analysis, economic or financial resource impacts fall outside the REPA scope. From a decision-making point of view that affects the company's bottom line, however, this last impact category may have trumped all other considerations in the 1969 Coca-Cola study. Once profit is no longer a mitigating consideration, statements regarding ecological superiority of one alternative over another seem clearly supportable. One conclusion from the 1974 study reads:

> Substantial improvements in one-way containers would be needed to increase their ranks to a tie with 10 [trip returnable bottle]; on the average, the impacts of the second ranked container would have to be cut in half to equal 10 [trip returnable bottle]. It is unlikely that the overall impact profile of any container will be improved to match or surpass that of 10 [trip returnable bottle] in the near future. (EPA 1974)

The report concludes that, maintaining a system to support bottle return for refilling is the clear, objective, environmentally beneficial, and responsible choice. Refilling a bottle ten times is twice as good as the next best alternative. Comparing these two similar studies carried out by the same consulting research company, it is possible to infer nefarious motives on the part of Coca-Cola to increase the use of plastic bottles instead of investing more heavily in the reusable glass bottles. Keeping the quantitative details of the original REPA secret for half a century gives the impression that there was and may still be something, in the approach or in the conclusions, to hide.

Perhaps more interesting, however, is the decision to include the financial impacts category in the first study. Considering that "profit" would later come to be known as one of the three primary "triple-bottom-line" elements (Elkington 1998), the 1969 study proves to be prescient in overtly considering not only "planet" but also "profit" impacts as two central metrics to be *balanced* in order to achieve a truly sustainable future. Had the study attempted to also consider how to measure *social impacts*, or the third bottom line concerning "people," it would have been almost 20 years ahead of its formal succinct articulation by the World Commission on Environment and Development. Coca-Cola would have been an example of sustainable industrial development that works to "meet the needs of the present without compromising the ability of future generations to meet their own needs" (Bruntland 1987). It is not useful to note that this private study did not consider one missing formal element that would take a future UN-commissioned report to develop.[3] Given the growing acceptance of this triple-bottom-line approach by a greater proportion of society, and since the study was limited to a detailed evaluation of the packaging and not the contents, it may actually help Coca-Cola's long-term future corporate image to finally make the details of this seminal document public. It is important to note that the REPA focused only on packaging and distributions and that the impacts of growing, harvesting, processing, and combining the raw ingredients into the elixir that has been key to Coke's global dominance are excluded from the study. Sharing a 50-year-old REPA does not risk divulging even a single ingredient making up the secret recipe of what is in the bottle.

4.4.3 The Record Reveals Clear Intentions

During his address to the Technology Assessment Board in hearings held by the US Congressional Office of Technology Assessment, in June of (OTA 1976), Harry Teasley was quite transparent about many of the study's key details and conclusions in comments and answers to follow-up questions posed by the largely sympathetic and supportive board, staff, and advisory council members in attendance. Key parts of the original conversation outline a familiar dilemma regarding environmental degradation and humankind's attempts to mitigate them at the source. Teasley also frames technical conclusions from the 1969 REPA as well as any other net environmental impact studies and, even more generally, the broader use of data and information from technology assessment activities within a four point conceptual framework:

*The first [concept] deals with acute public health or environmental issues...*activities, operations, or products that have the potential to cause acute public health or environmental problems should be controlled, related, or prohibited...The objective of this management philosophy, or this concept, is to prevent disasters.
The second concept deals with the short-to-medium-term use of the "commons" – air, water, land, resources... When the aggregate use of the "commons" begins

---

[3]The third triple-bottom-line assessment metric attempts to quantify good or ill effects of a system on "people." Just a few years after the REPA was concluded, the social would also emerge as a central but general concept in the contemporaneous systems thinking, *Limits to Growth*, report published by the Club of Rome (Meadows 1972).

to approach their natural carrying capacity, adverse impacts begin to occur. These impacts are costs to society. Products and services should include all costs, direct, environmental and social, in their cost structure. Therefore, the externalities should be internalized by setting limits via standards, or by charging direct fees…

The third concept is a different management concept. It deals with the use and allocation of resources… Over and above the management implied in the first and second concepts, society, acting through governmental institutions, should allocate private resources by managing the cost, availability, or terms of sale, for products and services within the economy… Implicit in this concept is the view that society can best handle the allocation process and make determinations on what products should exist and what products should not exist.

The fourth concept deals with the long-term use and availability of resources, and societal value systems relating to growth, consumption, and life-style. It can be stated in the following manner: Over and above the management implied by the first and second concepts, society, acting through governmental restitutions, should control the overall use of resources, and search for a no-growth equilibrium economic system. In other words, put a cap on economic envelopment or resource utilization. The objectives of this concept are to reduce consumption and to take a longer term view of the world. Implicit in this concept is the belief that society should be culturally intensive rather than use intensive, capital intensive at the consumer level rather than flow intensive, and labor intensive in many sectors rather than energy intensive. (OTA 1976)

With these four concepts, he illustrates a range of considerations that influence the ultimate path a society can choose to better manage resources and reduce the negative environmental impacts of industrial activities. To Teasley's credit, this initial framework describes a complex and conflicting set of goals that are not easily resolved, which Teasley is not afraid to state publicly in those early days of the modern environmental movement.

Practically speaking, however, Coca-Cola took on a challenge posed by the very consumer market in which they were already a dominant player in order to address negative environmental impacts of their products with a fact-based study. Teasley summarizes the original reason for taking the analytical approach to address a perceived threat to that dominance:

> What we saw happening to us is that, our world was changing and we were getting criticized at times for the direction that it was taking. We had to understand not only the economics that were bringing about that change, but also the environmental and market impacts that were associated with that change. Our studies were simply to provide management with an additional tool that they didn't have before. With that tool we could make R&D judgments about whether we ought to pursue a certain kind of development or not, and we could make procurement decisions. (OTA 1976)

He goes on to note the attendant benefits inherent in such an approach from a purely operational efficiency and business point of view:

From a business standpoint it has also been extremely useful to us because now that we know how much energy or how much diesel oil or how much gasoline is involved in a specific option, we can make long-term plans about what is going to happen to the cost structure of that option vis-a-vis another option. So we have improved our planning capability substantially by developing that data base and that analytical tool. (OTA 1976)

In the follow-up question-and-answer period on that spring morning in 1976, Teasley clarified a key point regarding the ultimate decision to use plastic instead of glass bottles that may have stayed locked in the original unpublished and detailed analysis and conclusion. There was a question from OTA's National Advisory Council representative, Mr. J.M. Leathers, who asked: "In the switch from returnables to the plastic bottle, did you make a net energy analysis to see where the break-even on energy would be?"

Teasley answered:

We have not switched. The plastic bottle that we have introduced in the marketplace did not substitute for returnable bottles; it substituted for already existing one-way glass containers. A net energy analysis on that move indicated [that], in the size range that we introduced which was a 32-ounce size, we were equal to glass in energy consumption. So it was a washout. Probably the assumptions and error one way or the other would tell us which one was really the lower energy consumer. We had determined that we had improved safety factors, and that we had very high market and consumer acceptance.... (OTA 1976)

So Coca-Cola introduced single-use plastic bottles in the early 1970s not because they were environmentally superior to their returnable glass counterparts but because they were roughly equal to the single-use glass bottles they were also using and they were perceived to be safer. Teasley does not go on to specify for whom or how plastic is safer. Since he does not tie the improvement to one or more of the study's formal impact categories, he may be referring to bottle breakability, which was not a category in the REPA. Perhaps most importantly, however, was that "high market and consumer acceptance" paved the way forward for plastic.

When asked again about the break-even point between the two system alternatives, Teasley did not hedge when he reported, "On balance, break-evens occur somewhere between 3 and 5 trips" for returnable bottles. After that, especially in light of the results from the subsequent EPA study, we can assume that the environmental impact is further reduced with every additional refill.

4.4.4 The Results

As Harry Teasley was testifying on technology assessment, he referred to the importance of net energy analysis as a driver for the evaluation of net environmental impact in production. He even referred to two books by the ecologist, Howard Odum, he had read to emphasize the emerging trend in industrial activities to embrace net energy "input-output analysis in economic terms and general system analysis" (OTA 1976). This sophisticated scientific approach seems, however, not to have adequately anticipated the actual effects of the sheer number of single-use plastic bottles on the environment over time.

Today, plastic Coke bottles can be found nearly everywhere. From North American supermarket isles, to Andean town corner stores, to European cafes and Asian restaurants, to movie theaters throughout the globe. Even the most exotic beach resorts will offer any number of internationally recognized brands of soft drinks. Unfortunately, and as we saw in Chap. 3, one tragic fact of the industrial world is that these same beaches are also destinations for millions of millions of tons of ocean-born plastic waste. The message in every bottle as they wash up on the shores of every continent is not a separate note romantically carried inside. Rather, the message emerges in the aggregate; with each year and each ton of flotsam that washes up or spirals endlessly in ocean gyres, the message is more pronounced. These plastic bottles, whether intact or torn apart by entropy into smaller and smaller particles, are the physical evidence of our collective design failure to safeguard our own home.

4.5   Three Basic "Rs"

"Reduce, Reuse, Recycle." The most popular dictum to have emerged from the green movement reflects a principle that, for decades, has been accepted as reasonable and actionable by billions of people. The slogan itself may have crystalized sometime after the first Earth Day, half a century ago, but the behaviors it describes and codifies have been used either separately or together for millennia. Whenever durable materials like stone or brick were carefully and skillfully salvaged and repurposed for new structures or even when patiently saved scraps of cloth served as feedstock to make quilts, value was recognized and material utility extended. It is the mantra for millions of individual consumers today who, concerned about how their own personal actions directly affect the course of future generations, carefully consider purchasing choices. It echoes in the ears of anyone who pauses to consider into which bin to drop some piece of plastic, the most ubiquitous, durable, and increasingly problematic packaging material, after searching for the sometimes nearly invisible "resin identification code."

Under *ideal* conditions, habitually applying one or more of the three "Rs" in the daily lives of the 7 billion plus people alive today can have a significant effect on what, how, and how much we collectively take in and give back during our time on the planet. If each person and corporation were to individually and consistently act according to these tenets and perfectly closed industrial recycling systems operated flawlessly, then effective, self-coordinated, global-scale resource management would be possible. The US Congress passed the Resource Conservation and Recovery Act in October of 1976 (US EPA, 2013) through which the three Rs gained popularity. Since then, however, the total amount of resources being transformed has increased year over year. In spite of the aspirational goals set by the act, while individuals may have marginally reduced their own domestic consumption, there has been no overall net national reduction, and the annual rate of resource transformation has tripled globally since then (UNEP 2016). Furthermore, neither consumer or corporate behavior nor the industrial systems needed to create a closed loop to approach a zero material waste economy have manifested.

4.6 Conclusion

This first attempt by the Coca-Cola Company in 1969 at quantifying specific impacts that various industrial processes and consumer behavior patterns have on the environment represented a helpful start. It reflected emerging trends in Europe and other parts of the world that would eventually become codified as an international standard. A full two decades after Harry Teasley testified to the US Congressional Office of Technology Assessment in June of 1976, the International Organization for Standardization would publish the first standardized principles and framework for life cycle assessment to support environmental management.

The work carried out by a global network of hundreds of dedicated scientists, economists, engineers, and industrialists during 20 years leading up to that time and the more than 40 years since have built a new internationally recognized scientific field called life cycle assessment. The discipline continues to expand and attract attention from more people that do not shy away from complex questions. These questions attempt to make sense of the myriad subtle interactions between elements, compounds, industrial processes, and consumer behavior throughout a product's life.

LCA analytical techniques represent a step forward for everyone concerned with measuring the ultimate impacts of human activity. What good are sophisticated techniques, however, if they ultimately fail to provide actionable insight to decision-makers? The next chapter discusses LCA details and how they create a picture of system impacts to guide rational decisions on what and how to make.

References

Brady LM (2014) The Department of Defense and its precursors: history, responsibilities and policies. In: Fairfax SK, Russell E (eds) CQ Press guide to U.S. environmental policy. CQ Press an imprint of SAGE Publications, Washington, DC, Inc, pp 255–268

Brundtland GH et al (1987) Brundtland: our common future (report for the world... - Google Scholar). Oxford University Press, Oxford

Elkington J (1998) Cannibals with Forks: the triple bottom line of 21st century business. New Society Publishers, Gabriola Island/Stony Creek

Hunt RG, Franklin WE, Welch RO, et al (1974) Resource and environmental profile analysis of nine beverage container alternatives: final report

Kingman AZ Airplane Boneyard. https://airplaneboneyards.com/kingman-arizona-airplane-boneyard-storage.htm. Accessed 28 Jun 2020

Marcy S (1980) Generals over the White House. Appendix A: Eisenhower's 1946 Memo on
    "Scientific and Technological Resources as Military Assets." WW Pub, Atlanta https://www.
    workers.org/marcy/cd/samgen/genover/pcnvrt06.htm
Meadows D H (1972) Limits to growth by Meadows, Donella H. (October 1, 1972) Mass Market
    Paperback
Novick D (1959) The federal budget as an indicator of government intentions and the implications
    of intentions I RAND
Plastics Institute of America Inc. (1991) Plastics in food packaging conference: plastics in food
    packaging: proceedings of the 8th annual Foodlas conference
Jardini (1996) A brief history of the RAND Corporation. https://www.rand.org/about/history/a-
    brief-history-of-rand.html. Accessed 2 Feb 2018
Technology Assessment Board (1976) Technology Assessment Activities in the Industrial,
    Academic, and Governmental Communities NTIS order #PB-273435 - Hearings before the
    Technology Assessment Board of the Office of Technology Assessment Congress of the United
    States - Ninety-Fourth Congress Second Session - June 8, 9, 10, and 14, pp 105–116
Toepfer G Bioconcepts the origin and definition of biological concepts a multilingual database.
    http://www.biological-concepts.com/views/search.php?term=814&listed=y.   Accessed   23
    Jan 2018
Truman HS (1945) Harry S. Truman: special message to the congress recommending the establish-
    ment of a Department of National Defense. In: The American Presidency Project. http://www.
    presidency.ucsb.edu/ws/index.php?pid=12259. Accessed 21 Jan 2018
UNEP (2016) Global material flows and resource productivity. In: Schandl H, Fischer-Kowalski M,
    West J, Giljum S, Dittrich M, Eisenmenger N, Geschke A, Lieber M, Wieland HP, Schaffartzik
    A, Krausmann F, Gierlinger S, Hosking K, Lenzen M, Tanikawa H, Miatto A, Fishman T (eds)
    An Assessment Study of the UNEP International Resource Panel. United Nations Environment
    Programme, Paris
US EPA O (2013) EPA history: Resource Conservation and Recovery Act. In: US EPA.
    https://www.epa.gov/history/epa-history-resource-conservation-and-recovery-act.   Accessed
    27 Jun 2020
Woods R, Petersen C (1995) World War II and the birth of modern recycling. Waste Age 26:226–238

Chapter 5
Life Cycle Assessment

Abstract Born as a military accounting tool and adopted by corporate America to increase efficiency, life cycle assessment has matured into an indispensable method for environmentalists to quantify ecological impacts.

Chapter 4 presented how in the 1950s the US Military employed a life cycle approach to optimize resource allocation by applying scientific scrutiny to the budgeting process. In the 1960s, Coca-Cola was the first corporation to implement similar techniques to include the consideration of environmental impacts as part of their strategic financial decision-making. Nearly 30 years later, the International Organization for Standardization (ISO) codified such techniques as life cycle assessment. The ISO published a concise definition of this complex and sprawling activity in the first formal global LCA standard in 1997.

Environmental LCA, which takes a life cycle view of a product or activity from gathering raw materials through end of life, is now a recognized formal scientific discipline and currently the purview of scientists, economists, and accountants. Specialized terms and methods, fundamental to the proper assessments carried out by these technical disciplines, continue to challenge LCA's broad acceptance by the design community. Design professionals familiar with origins and fundamentals of LCA, however, can more thoroughly engage with these processes to actively shape a design's ecological profile and provide additional value to their clients and the public. A working knowledge of the basics benefit the non-specialist. This chapter will focus on "goal and scope," the first of the four primary LCA phases as an introduction to the three others that will be covered in later chapters.

5.1 Proto-LCA to Standardized Practice

Two decades passed between when MRI conducted the first REPA for the Coca-Cola Company and the development of the first standardized LCA guidelines by the Society of Environmental Toxicology and Chemistry (SETAC) in 1993. Walter Klöpffer presents a thorough account of the transformation during the intervening years from the earliest "proto-LCA" or ecobalance studies in the early 1970s to what are, today, widely accepted scientific LCA standards. During this time, fossil fuels, waste disposal, and packaging dominated the studies. By the 1980s, a growing number of companies were misusing proto-LCA studies to support public

© Springer Nature Switzerland AG 2021
J. Cays, *An Environmental Life Cycle Approach to Design*,
https://doi.org/10.1007/978-3-030-63802-3_5

facing comparative product assessments to gain market advantage. SETAC's new guidelines, developed in conjunction with European public research institutes, corporate partners, and early LCA consultancies, provided a more rigorous methodological framework to counter this trend (Klöpffer 2006).

European academic researchers made important methodological strides in the early 1990s and SETAC's 1993 publication *Code of Practice* represented the first formal consensus expert opinion on the state of the practice and included the following main sections:

LCA Definition
Technical Framework for LCA
Data Quality
LCA Applications and Limitations
Presentation and Communication
Peer Review
Future Research Needs

SETAC continued to form steering committees and working groups made of LCA practitioners and experts to discuss and make further recommendations on:

Improvement analysis
Impact assessment and especially the integration with other techniques
Allocation
Development of an educational module and training course
Screening and streamlined LCAs
International databases
Practical applications
SETAC (1994)

As the 1990s progressed, these working groups advanced progress in many technical areas and sharpened methodological approaches and standards that would be handed off to the International Organization for Standardization. Industrial researchers were well represented from the beginning. Dr. D. Postlethwaite of Unilever Research chaired SETAC's LCA Steering Committee and was available to answer inquiries regarding activities within a growing LCA community.

Important to emphasize is that SETAC was and is a professional organization not a regulatory body. It has no authority to set and regulate standards. The Society was established in 1979 in the US to promote the use of a multi-disciplinary collaboration to solve problems caused by the impact of chemicals and technology on the environment. It continues to be dedicated to the study, analysis and solution of environmental problems, the management and regulation of natural resources, research and development and environmental education.

From 1994 after SETAC's publication of the *Code of Practice*, the International Organization for Standardization began developing and publishing widely recognized and agreed upon standardized protocols. From 1997 to the latest published 2006 LCA standards 14040 and 14044, the ISO gave LCA the necessary formal authority it needed to penetrate deeper into industry, academia, and regulatory bodies worldwide. In 1997, the ISO integrated one of SETAC's original phases, "life cycle improvement assessment," into the entire LCA methodology and added the "interpretation" phase as the fourth (Rebitzer et al. 2004).

This chapter focuses on the first of the four iterative LCA phases established in 1997 and continued in the ISO 2006 edition of 14040: Environmental management—Life cycle assessment—Principles and framework. The iterative nature of the LCA process across phases allows the discussion of this first "goal and scope" phase that precedes the subsequent three phases presented in Chaps. 6, 7, 8, and 9 as they potentially augment and enhance evolving design workflows.

5.2   What Is an LCA for and What Exactly Does It Measure?

"LCA is a technique for assessing the environmental aspects and potential impacts associated with a product by: compiling an inventory of relevant inputs and outputs of a product system; evaluating the potential environmental impacts associated with those inputs and outputs; [and] interpreting the results of the inventory analysis and impact assessment phases in relation to the objectives of the study." ISO 14040: Environmental management—Life cycle assessment—Principles and framework 1997(E). p.ii; International Organization for Standardization: Switzerland.

LCA documents are ultimately intended to produce conclusions regarding *potential* environmental impacts. In one sense, "potential" refers to future impacts that may or may not be realized depending on certain contingent conditions. In another sense, "potential" implies the likelihood of stored capacity embodied in the assessed good or service to impact the environment. This conclusion is based on the assumption that there is a set of normal, constant conditions under which the odds are very high that the end result in the real world will match the modeled prediction.

The datasets regarding conditions, materials, and use patterns on which the LCA conclusion is based are a critical part of an LCA report's stated assumptions. These sets may be more or less complete or correct and can materially affect the conclusions regarding future environmental impact projections.

If a company wants to know how what it makes will potentially impact the environment throughout its life cycle, there are five typical key impact categories that get measured and documented. After assessing all available, relevant, and compliant data, a numerical value is assigned to a product or service that represents its potential to:

1. Create acid rain (acidification)
2. Kill streams rivers and oceans (eutrophication)
3. Make smog (tropospheric ozone)
4. Make the hole in the ozone layer bigger (stratospheric ozone)
5. Make the earth warmer (CO_2)

These are the five primary impact categories, but they are not the only things that can be with respect to potential environmental impacts. Depending on the method followed, the primary list can be expanded to include the potential to:

6. Increase risks to human health
7. Reduce the amount of usable land available to all species
8. Decrease the amount of usable water
9. Poison the environment with toxic chemicals
10. Deplete abiotic resources including fossil fuels from not renewable deposits

Abiotic depletion potential actually spans multiple impact categories since there are many different materials included and varying time horizons of resource renewal. Comparing the depletion impacts of many different materials, with a wide range of practical implications when they become unavailable as a resource, makes aggregation of effects expressed in one metric difficult to establish (van Oers et al. 2002).

Done right, a life cycle assessment produces a technical and comprehensive document that accurately compiles and presents detailed, copious, and correct data on what gets taken from—"inputs"—and given back—"outputs"—to the earth, along with a projection of the potential harm or benefit it will have on the environment. The LCA report is, ideally, a neutral document that clearly states all assumptions and qualifies its conclusions.

5.2.1   Two Conjoined Standards: ISO 14040 and ISO 14044

ISO 14044:2006E Environmental management—Life cycle assessment— Requirements and guidelines is a foundational guiding document for LCA practitioners that clearly and succinctly cover the elements of an LCA. Its primary reference document is *ISO 14040:2006E Environmental management—Life cycle assessment—Principles and framework.* Taken together they describe the primary considerations needed to conduct and report on formal LCA studies. These two documents, and subsequent standards that refer to them, provide answers to many individual challenges raised by those committed to improving how we produce and consume. The scientific techniques underpinning proper LCA activities create stronger defensible positions for any sustainable design claims. Understanding the information in these documents is especially useful to designers looking for rational support for individual sustainable design decisions. Any effective environmentally conscious organized group of design professionals building a comprehensive guidance system from scratch would likely agree on many of the principles, frameworks, requirements, and guidelines the scientific community settled as they searched for answers to fundamental questions. Those questions center on how to more accurately quantify the damage caused by human activities with the ultimate goal of mitigating and reversing it over time.

The questions are not technical although the techniques used to systematically answer them can be and require some explanation. These practices, taken together, follow a higher standard of care to more accurately assess the environmental impacts created in any system. Using LCA techniques can enhance the design process. They sharpen thinking and provide firm grounding to make the case for one design decision over another based on the major features of the four iterative phases of any proper life cycle assessment. The four phases are:

Goal and scope definition
Inventory analysis
Impact assessment
Interpretation

The following are questions asked and answered by these two foundational ISO documents. The main difference between life cycle assessment and design activities is the obvious focus on deeper analysis in LCA vs. synthesis in design. There are, however, numerous general similarities as well, such as iteration as best practice in both LCA and design. The following presents the two activities through a series of questions and answers while identifying how LCA can directly inform the quantitative environmental profile of a design. Three kinds of design products of different sizes, complexity, and durability illustrate how these LCA phases define key aspects that affect environmental impact. Footwear and beverage containers represent relatively simply defined systems. Buildings are used to discuss the challenges with more complex ones. Digitally delivered services are a different kind of system that manifests virtually whose environmental impacts are at once easier to quantify since electricity is the primary medium yet complicated by the many intermediary devices and networks required for their dissemination and consumption.

5.3   Goal and Scope Definition

All key features of a study are initially established in this first LCA stage. It is a rigorous proto-LCA in miniature purposely lacking many specific details on which a proper full study depends. The supporting rationales and documentation, regarding what and whom the study is for and how it will be carried out, create a first sketch of a proposed design's relative net ecological burden or regenerative potential. It defines the requirements for the subsequent and more detailed, data inventory and environmental impact stages. Early review, evaluation, and interpretation of what is and is not appropriate or necessary to meet a study's goal are initially carried out before proceeding with detailed investigations in later stages. So all LCA phases are represented in the goal and scope phase as recursive fractal seeds that will be modified and developed into the more comprehensive LCA as the study progresses.

5.3.1 Goal

At the start of most design projects, it is critical to identify the purpose and intended group who will benefit from the completed project, product, or service. Without this, the design remains in the purely formal or speculative realm disconnected from the public or a particular consumer or beneficiary. Similarly, LCA practitioners must answer two practical questions related to the study's overall goals and intended audience to properly structure their work.

The first is question the practitioner must answer is, *"What is the purpose of the study?"* Life cycle assessments can serve many purposes, and the facts gathered and formatted are directed to a particular end. The study could benefit new product development or the improvement of an existing one through revealing environmental "hot spots" and evaluating alternatives in the production process or any other life cycle phase. Designers can use LCA to test and evaluate assumptions about individual "sustainable design" elements. It can inform strategic planning within a private or public organization. Broad environmental public policy decisions can use LCA conclusions as supporting evidence. Companies may use LCAs to support environmental performance claims through third-party-verified environmental product declarations (EPDs). These are discussed in detail below. Before starting any LCA activities, as with any other activities that will demand time and effort, it is important to be clear on its ultimate purpose or goal.

The second question clarifying main goals of the study is *"For whom is it intended?"* In initial discussions, the focus of the study may shift to target particular issues important to the sponsoring key stakeholders or those who eventually will be affected by the conclusions of the LCA. The public's interests, for example, are not the same as those of a management team charged with bringing a product to market. A study that will be used as the foundation for a third-party-verified EPD, for example, will likely be more thorough and more complex than one carried out to provide insight into internal manufacturing or material extraction operations. The results of the latter type are not usually intended for sharing with anyone outside of a small group technical team experts. A study used by designers to validate design decisions to a client has both different purposes and data quality requirements than one intended to be shared with the general public.

Final communication structure and form and the level of detail included depend on both the study's purpose and audience. Clearly identifying both at the start of any LCA activity guides the work and report of findings. Once the overall goal is established, the next step is defining the study's scope.

5.3.2   Scope: What Is the Product or Service System and What Is Its Function?

The scope of any LCA is determined by numerous elements. Answers to the first questions regarding *what the product or service system is and what is its function* can seem simple and direct in the abstract but can become complicated as complexity of the system increases or multiple secondary functions are considered. These are questions that design professionals routinely ask but for different reasons and from different perspectives. The holistic adage "form follows function" coined by the architect, Louis Sullivan (1896), is still applied today from building design, to biophilic product design, to software development. The value, applicability, and truth of this phrase have been debated throughout the twentieth century but have survived in common usage or misusage to conjoin two concepts. Many designers consider it a tired cliché that struggles to keep pace in an age of digital interfaces or black-box information age processes. The analytical purpose of two similar but clearly separate questions used to scope LCAs may provide less loaded ways to consider what something is intended to *be* (form) and *do* (function). In the scoping phase, these practical components are initially established as propositions open to later revision in an iterative process.

Take, for example, a pair of shoes. Their primary purpose is to protect feet from injury while walking, and, for this, a pair of sandals or flip flops satisfy the requirement as shoes. Secondary or more specialized benefits that modify how they are expected to perform can modify the system definition and function. Boots built for wet or cold weather must do more than protect the soles of the feet from injury. They must keep feet warm and dry and may provide traction on slick surfaces. Athletic shoes address many different external conditions and are expected to support and even enhance athletic performance during an event. Shoes can also be optimized to express social status with little difference in physical performance.

Secondary functions provide useful specificity regarding the subcategory of system under consideration. In this simple example, boots and flip flops are not comparable footwear systems (Jolliet et al. 2015). Depending on the purpose of the study, especially when it will be used to compare two or more options, a secondary system function must move into a primary position to describe fundamental traits material to a proper "apples to apples" comparison. A concise description for a footwear system, valid for making useful comparisons, would be "one pair of subzero weather winter boots."

In more complex systems, comparability between systems becomes significantly more challenging. Buildings, for example, represent highly complex, systems of interconnected systems. Individual components, from concrete or steel structural frames to the interior finishes, work together to define a building. Whole building LCA (WBLCA) presents numerous scoping challenges. Even within a single building type, there are many situational and site-based qualifiers that make direct comparisons difficult. Recent proposals to establish a formal WBLCA goal and scope taxonomy move to address the well-documented challenges in establishing general standards in this environmentally important sector (Rodriguez et al. 2019). Standardizing basic project scope information dealing improves comparability between studies. Consistent parameters regarding building scale and performance include the building footprint size and gross square foot floor area. Additional geographic and site-specific characteristics such as soil class as well as climate and seismic zone enable meaningful comparisons.

Physically materialized systems such as shoes or buildings are not the only design objects of study requiring the same scope specificity. Virtualized services and experiences depend on physical infrastructures and interfaces to distribute electricity and information. Different digital content requires greater or lesser material and energy resources to create, distribute, and consume. The negative impacts caused by waste byproducts are largely determined by grid energy mix. Broadband-delivered virtualized cloud-based services reduce local hardware needed to play a video game. Cloud servers, however, consume energy from a grid that may still be partially or completely powered by coal-fired electrical plants creating the same environmental impact dilemma as the coal/electric car. It considering alternatives in a "dematerializing economy," the real question may be what is the environmental impact of a digital version of a good vs. its physical counterpart. The ecological costs associated with digital dissemination of a book read or image viewed on a smart device can be compared with print versions. This would prominently figure into the scoping exercise.

5.3.3   How Specific Is Specific Enough?

What is and what is not included in a system is a critical part of the scoping phase. Valid comparisons require a single consistent unit used in various scenarios. The *functional unit* establishes a measurable and quantifiable reference to make comparisons of various alternative scenarios. The soft drink industry provides the classic example of a functional unit (FU) as a volume of 8 oz or 300 ml of soda delivered to a single consumer in an open or closed container. This measurable unit can be used to compare various container alternatives such as glass, aluminum, or plastic with all the attendant process and distribution variables associated with each to perform its intended end function—someone enjoys a single cold carbonated drink. Immediately prior to imbibing, this drink can be poured into a paper cup at a fast food restaurant, drop out of a vending machine in a bottle or can, removed from a domestic kitchen refrigerator, or arrive in an insulated cooler far from any mechanical refrigeration.

Packaging is a fairly discreet and relatively easily defined system to design and analyze alternatives. In a more complex system such as a building, the total volume or mass of materials does not capture an assembly's qualitative and performative aspects needed to compare it to another that would provide the same service over time. To serve as a functional unit, the declared unit expressed in simple unqualified terms such as a 100 m² of a solid external building envelope material like wood or brick must include more descriptive detail of how it is expected to perform over time. Details such as required useful life span of 50 years and UV light, water, fire, and impact protection make comparisons between alternatives possible (Simonen 2014). In a building, these performance specifications would apply to each defined building subsystem such as roofing, glazing, flooring, finishes, mechanical equipment, telecommunication infrastructure, or insulation.

For the same reason that defining the project scope in a WBLCA study is difficult, defining a functional unit for meaningful comparisons between studies is nearly impossible given all of the possible variations in building construction, performance, and program. Much like the subsystems that make up a building, defining the functional unit of a building requires both a unit of measurement (i.e., m^2) in addition to an expected service life of the building in addition to succinct quantitative and qualitative descriptions of the building's performance and ultimate purpose.

All products and services, including digitally delivered ephemeral ones, have physical components that contribute to environmental impacts. The functional unit of a digital product may be defined like that of a drink delivered through either a bottle or can since it usually can be defined as a discreet creative work unit like a static or moving graphic, e-book, movie, song, or video game expressed through a particular file format like .JPG, .MP4, PDF, or any of hundreds of proprietary formats. The sequence of bits conveying information is typically delivered through vast intra-/intertelecom networks and mediated for final consumption on an array of physical electronic devices whose negative environmental impacts may or may not be included in the same LCA.

5.3.4   What's In and What's Out?

The scoping phase starts by defining what a system is and what it does. The complex and interconnected nature of each thing next requires clear definition of the *system boundary.*

A percentage of the energy, raw materials, and factory solid, liquid, and air waste emissions released in producing each designed object can be assigned to a single, boot, brick, book, or byte. The associated impacts of each of individual good or service throughout and after its useful life can also be tallied. These will be extremely small fractions of the total amounts of the eco-/technosphere matter and energy exchange, but there is a theoretical exact amount.

But what about the myriad secondary, tertiary, and even more indirect inputs and outputs without which the final system under consideration cannot exist? Any process or material input or output that does not contribute a significant percentage to the final assessed environmental impact lies outside the system boundary and is excluded from the study. That means the impacts associated with the manufacture of the truck that carries a load of timber off the side of a mountain would lie outside the system boundary while the emissions from the truck as it makes its trip to the saw mill fall within it.

A flow diagram shows the relationships between inputs and the intermediary components in the processes that produce outputs. These diagrams feature bounding perimeters. Parts of the processes used to produce the system are inside, and the adjacent parts that fall outside indicate the extent of the assessment. Essentially what is in and what's out determines whether the LCA is a cradle to grave/cradle, cradle to gate, gate to gate, or gate to grave/cradle study.

Unlike vast regions of terra incognita marked at the edges of ancient maps with warnings to navigators that "here be dragons," the parts of the process that fall immediately outside the system boundary are typically knowable. The data and modeling needed to describe them are purposely excluded if they are simply not part of the stated object of the study. Initial boundaries can be iteratively expanded or contracted at a later point as required by an LCA's evolving goals and scope.

5.3.5   How Little of Something Is Too Little to Include?

Cutoff criteria set the quantitative threshold for what is included and excluded in the study. As unit processes and product systems and subsystems are evaluated, the associated material and energy flows or environmental impacts that fall below the established threshold are purposefully left out. Without cutoff criteria, the interconnectedness of everything quickly overwhelms. Recognizing that, for example, delivering a can of soda to a consumer requires an array of machinery to make, fill, transport, and track the can and those machines were built by machines which were built by other machines and so on back to the start of the first industrial revolution, it is reasonable that a large number of these subprocesses must be excluded from any calculations. The impacts from these secondary, tertiary, and lower-order systems typically fall below the cutoff criteria threshold since they are not easily quantified (due to a lack of reliable data about them) nor do they contribute a material percentage to the ultimate system impact assessments.

LCA experts acknowledge an apparent paradox: accurately setting appropriate cutoff criteria in a study requires the results provided only by a full and complete study. The chicken and egg problem is typically addressed through first taking a rough cutoff for mass, energy, or cost flows below some stated fraction of the total. Appropriate estimates range from a high 1% (Curran 2012) to an order of magnitude lower 0.1% (Hauschild et al. 2018)—but with the caveat that high environmental impacts can be caused by miniscule quantities of some substances that may fall below a cutoff threshold. LCA's iterative nature allows for the inclusion or exclusion of smaller quantities of certain high-impact elements or compounds. Sensitivity analyses and other statistical evaluation methods refine the criteria throughout the LCA process. Examples of low-quantity high-impact substances include radioactive isotopes, toxic chemicals, or rare metals mined and processed using environmentally hazardous methods. While nonexistent in the soda can example, these materials may be critical in the products or components in which they are used and be difficult if not impossible to substitute in a design. Iteratively qualifying their appropriate position inside or out of the system boundary requires establishing a quantified relationship to cutoff criteria based on their potential to do harm at some point in their life cycle.

5.3.6 What if the Process Used to Make One Thing Simultaneously Produces Other Things?

Most designers are commissioned to design a single object or system. As challenging as it can be to establish the system's boundary conditions and cutoff criteria in a design, the next step is arguably even more so. Many industrial manufacturing processes produce more than a single useful product. The various inputs and outputs of the entire process may need to be divided and separately assigned or *allocated* to the different resulting products. For instance, making a wooden table involves subtractive processes, to shape its individual elements from larger pieces of dried lumber, where the leftover wood has other potential uses. Depending on the specific processes used (sawing, coping, milling, drilling) and the wood feedstock used (hardwood, softwood, solid, laminated, fiberboard, etc.), useful wooden *co-products* could include solid wooden toys or sawdust to be used as economically valuable feedstock to make new pressed fiberboards, wood pellets, or for its direct use as fuel in an onsite furnace.

Appropriately allocating percentages of environmental resources and waste emissions among multiple-designed primary products and co-products can be a controversial activity. There is disagreement around the general methods used to make decisions around how specific percentages are split and assigned since the results can vary across methods or some are not practical using present data and software (Curran 2012). There are a few different approaches to allocate or to avoid the need to allocate. For those analyzing primary agricultural or industrial products (wheat, energy, steel, lumber, etc.), it is critical to establish a clear and defensible allocation strategy since the co-products are the feedstock for entire subindustries (straw, slag, cellulose, etc.) and can. For the designer focused on downstream systems, it is more important to grasp the broad implications of these various approaches than be able to technically resolve the problems inherent in tracking and allocating the environmental impacts along each flow path. Consideration of upstream natural resource extraction and processing systems may, however, provide insight into alternative design approaches.

ISO 14044 4.3.4 establishes a procedure to address allocation problems through a series of steps (ISO 2006b). Step 1 avoids allocation by either dividing unit processes into multiple subprocesses for each co-product and collecting input and output data related to each discreet subprocess (1a) *or* expanding the product system to include the additional utility/value the co-products provide (1b). 1a creates separate LCAs for each co-product that can be taken together to find the total flow of inputs from and outputs to the ecosphere. 1b produces a single LCA inside of which the inputs and outputs of the secondary co-products are calculated and subtracted from those of the primary product. In the simplified wooden table example, the utility a table provides is different than that of a toy or sawdust. The subprocesses not directly associated with the table are placed outside the system boundary *or* the table, and its co-products are grouped together as one expanded system that considers a set of wood products together with all additional functions they provide.

If neither of these approaches is appropriate to the system under study and allocation cannot be avoided, Step 2 partitions system inputs and outputs to reflect physical relationships such as each co-product's mass percentage relative to the total mass or each co-product's proportional financial market value relative to a total. If it is not possible to establish comparable physical or economic units, other value choices can be used as a last resort. Natural sciences provide the most reliable allocation approaches. Clearly documented links that exist between emissions, for example, relative to increasing or decreasing amounts of co-products create a firm basis for proportional allocation.

The closer a study's method for linking relationships between the objects of study is to the natural sciences, the more grounded it is in the natural world and the easier it is to express its conclusions in physical terms. The well-established economic proxies used in LCAs create reasonable and defensible equivalencies; but they are at least once removed from directly describing the physical impacts attributable to a physical process or system. It is, therefore, always preferable to allocate on a physical basis when causal links exist (Jolliet et al. 2015). This same preference hierarchy holds for designers and others reviewing LCA studies about various product system alternatives to support one choice over another.

5.3.7   *What Are a Design's Most Important Physical Impacts to the Environment, and How Are They Indicated and Modeled?*

Each of the assembled materials specified in a design will impact the environment. Life cycle impact assessment (LCIA) is one of the four major iterative stages in the overall LCA framework (ISO 14040 2006a) but also appears as an initial consideration in the goal and scope phase. Some materials used to execute a design in physical form will negatively affect some environmental impact categories more than others.

There are numerous impact categories to be considered and modelled (Jolliet et al. 2015), and some will be more relevant than others depending on the system being studied. For instance, all metals must be refined from ore and processed into billet and then shaped into finished goods. This high-temperature energy-intensive process emits carbon dioxide from burning fossil fuel. Wherever metals are used, greenhouse gas emissions will likely be a primary consideration. Higher percentages of greenhouse gas in the atmosphere increase the potential for global warming and the resulting negative climate change impacts (Masson-Delmotte et al. 2018). When specifying metals, it is likely that "climate change" will be an important environmental impact category to focus on. This is mainly due to the high temperatures used in the refining process requiring a high exergy fuel source, typically coal. "Global warming" may be the term used depending on which modeling system is used, but in either case radiative forcing, caused by the increasing concentration of CO_2 and other greenhouse gasses, is the universally accepted indicator.

Agricultural products specified in a design including wood, vegetable fibers, and natural dyes that depend on fertilizer and pesticides will likely consider the negative effects of nutrient loading on freshwater and marine ecosystems. In these cases, "eutrophication" is a likely impact category to include with several impact indicators including the relative amounts of dissolved oxygen, chlorophyll a, phosphates, and nitrates in streams, rivers, lakes, and oceans.

In these two impact category examples, climate change is a global impact, and eutrophication impacts are typically local or regional. The scale of the impacts is determined by how the indicator substances are released and travel into and throughout the ecosphere over time. The relationship between specific emissions, effluents, and solid wastes and their negative environmental impacts are represented through models that multiply the amounts of one substance by a characterization factor (CF) to produce a value that can be added to others in order to estimate the total impact. This can either be expressed as an attributional (midpoint) or consequential (endpoint) impact. A model that translates many contributing greenhouse gas emissions into single equivalency impact score would take gasses such as carbon dioxide (CO_2) x 1, methane (CH_4) x 28, nitrous oxide (N_2O) x 265, and sulfur hexafluoride (SF_6) x 23,500 and typically express them as total single midpoint characterizations score in kg CO_2e or kilograms of carbon dioxide *equivalents* over a certain time period. This example uses published 100-year global warming potential (GWP) factors for specific GHG emissions. There are different factors available for each gas's 20-year GWP time horizon (US EPA 2016). This points to the importance of which time element or duration is included in a study. Characterization factors are commonly used to model the various heterogeneous substances in all environmental impact categories, and LCA studies will clearly state which characterization models will be used. This will be discussed in more detail in the section covering life cycle impact assessment.

As important as the impact assessment phase is to a full LCA, depending on the study's overall goal and scope, it may or may not even be included. Most LCA-generated documents that designers would use, however, will include the impact assessment phase. The study may be limited to only focus on what the data inventory of material and energy that comprises a product system tells about it. Either way, the goal and scope will make explicit what major subsequent phases are included.

5.3.8 What If the Most Reliable, Accurate, and Precise Primary Data Are Not Available?

Since there are numerous calculated values on which a study is founded and impacts are assessed, an LCA model is only as good as the data on which it is based. Just as the key portions of the major "impact assessment" (LCIA) phase is included in the primary goal and scope definition phase, so too is included a preliminary "inventory analysis" (LCI). To those new to LCA, the subtle differences in the acronyms for these two main LCA phases with words beginning with the exact same letters can lead to initial confusion. A reliable impact assessment relies on an inventory analysis of data that is properly collected, calculated, and allocated. Only after final LCI results are assigned can impact category indicator results be calculated, so, in addition to being a constituent part of determining an LCA's initial goal and scope, the LCI is also an integral precursor to LCIA.

Heterogeneous LCI data is typical and perfectly acceptable for most LCAs. Documented primary readings taken from in-line or process sensors that measure compounds such as CO_2 or other indicator chemical emissions provide the most relevant and applicable data relating to that specific process. It is unrealistic to expect access to primary data on the energy, material, and emission flows for the thousands of unit processes that feed into a product system. A study based on 100% primary data today would be exceedingly time-consuming and prohibitively expensive. Secondary data, held by any of several public or private organizations, supplies most of the rest of the quantitative detail where primary data is absent. ISO standards make allowances for a variety of data:

> Such data may be collected from the production sites associated with the unit processes with the system boundary, or they may be obtained or calculated from other sources. In practice all data may include a mixture of measured, calculated, or estimated data. (ISO 14044 2006b).

The Swiss ecoinvent database, for example, has readily accessible process data on numerous sectors for use in an LCA. This data is typically averaged for each process and place. Eurocentric data is less accurate for processes carried out in North America or Asia that might, for example, each use a very different primary energy mix to power each unit process. Continent-/nation-specific databases, however, are growing and improving each year. As data reporting tools and requirements proliferate, data-aggregating organizations review, categorize, and incorporate new information sets to increase the utility and relevance of the services they provide (Matthews et al. 2014).

Generic LCI data is especially appropriate to use in the goal and scope phase as it can give a quick view of where system "hot spots" might be in order to focus on the right priorities and avoid spending time and effort later to collect detailed data on processes that don't contribute a significant portion of negative environmental impact in an LCA model.

Tools available to designers rely on these LCI data. The impact models they create draw upon the numerical values in the aggregated inventories. These values are assigned to various components within the geometrical simulations created in design software described in more detail in Chap. 8.

5.3.9 What Is the Standard Accuracy and Precision Level of LCI Data?

Data requirements are variable and depend on the study's stated goal and scope. The study's sponsors, along with those tasked with conducting it, set acceptable levels of required quality (primary, aggregated, uncertainty, precision, representativeness, etc.) appropriate to the study. Throughout the LCA's normal iterative steps, acceptable quality standards needed to adequately describe different parts of the product system may change.

Some parts of the system are considered as "foreground" and others "background." Foreground systems consist of *foreground processes* controlled by the decision-maker for whom the LCA is carried out. Background systems consist of *background processes* on which the decision-maker exercises little or no direct influence (Frischnecht et al. 1998).

The distinction between foreground and background process data creates a hierarchy for data collection. Generic data, common to similar product systems, requires relatively little time, effort, and expense to gather and integrate into the background system parts of an LCA. The more labor-intensive primary data collection effort can then be focused on the presumably smaller and more important foreground processes.

Designers evaluating and integrating LCA information into their workflow can use a similar hierarchy to discern products or components with an improved profile based on the differences in the foreground processes involved in their manufacture. Instead of treating all life cycle inventory elements as equally important, the focus is best concentrated on a presumably small number of unique parts of a product system that distinguish it from similar alternative systems. The choice between two distinct but equally durable and cost-effective fastening methods to make the wooden table in our example above, where all other components and processes are the same, may eventually hinge on what the data reveal about each fastening system. The background systems, wooden legs, and table top remain relatively constant, while the foreground show may show clear differences across one or more categories.

Inventory data describe important inputs and outputs to a product system. Among the wide range of inputs included are virgin or recycled primary mineral resources that comprise the product, energy supply mix, and necessary ancillary service and product inputs that fall within the system boundaries.

Primary data provide valuable numerical detail about individual process steps. Documented product system net outputs to air, water, and soil, not captured by pollution control devices, measured and recorded at the point of release create primary data points. A manufacturer may turn this type of information flow into a freely available continually published dataset or may choose to keep it for internal uses only. Not all US states require regular monitoring and reporting on point source discharges produced by industry and those that do concentrate on sectors and individual plants that have the potential to produce them over a certain threshold. New Jersey, for instance, sets thresholds for reporting on air emissions at or above 10 tons per year of volatile organic compounds (VOCs), 25 tons per year of nitrogen oxide (NOx), and 100 tons each per year of ammonia, carbon monoxide (CO), sulfur dioxide (SO_2), particulate matter 10 micrometers (PM10), or particulate matter 2.5 micrometers (PM2.5) (NJDEP 2018).

LCA professionals with access to a manufacturer's granular emissions, effluent, and solid waste inventory data can eventually evaluate and recommend improvements to foreground unit processes or subprocesses. Continuous improvement can be gauged against internal or external benchmarks, but all are driven by data quantity and quality.

In some cases, identifying and validating hotspots in LCI data and relating them to a functional unit's reference flows comprise a full study in its own respect. Straightforwardly called an LCI study ISO 14040 (2006a), it is generally distinguished from a full LCA by its lack of an impact assessment phase. LCI studies can provide designers and other decision-makers with insight into materials and processes with higher or lower recorded values over a range of inputs or outputs although, regardless of the data quality and quantity, they cannot be the sole determinant in making a final choice between two options. The goal and scope establishes initial data quantity and quality parameters and the need to go beyond an LCI to produce a complete LCA.

5.3.10   What Data Assumptions Apply to the Study?

Each study or model is an abstraction and simplification of a much more complex reality. Numerous foundational assumptions bracket and limit what the model can appropriately represent. This applies equally to the most complex global climate models as it does to a digital simulation of a small paper cup. Each LCA functional unit typically describes a material quantity, a specific purpose the unit will serve, and a service life period. Definitions and specifications of the materials and processes that make up the product system support qualified conclusions about a study especially when comparing it to other studies or when comparing product systems within a single LCA.

Some studies will focus only on a portion of the system's life cycle. This could be cradle to gate, a single gate to gate, gate to grave, and span multiple gates, or it could encompass the entire cradle to grave/cradle life cycle. Since many types and sources of data may be used, best practice requires qualifying assumption statements about information quality and breadth and its appropriateness to the life cycle segment under review.

ISO 14044 lays out ten general data requirements to support a study's goal and scope. Seven are qualitative, two are quantitative, and one has both qualitative and quantitative features. The qualitative requirements include three that describe breadth and depth regarding the data coverage parameters of *time*, *geography*, and *technology*; an overall assessment of how well these and other data *represent* the statistical population of interest from their categories; how *consistent* the methods for choosing and analyzing data are applied across all components that make up the study; how *reproducible* the results of the study are based on the kind of methodology and data value information provided; and what data sources are used. The two requirements expressed primarily in quantitative terms concern the *precision* or statistical variance of the data in each category and the *completeness* measured as a percentage of flow represented. Inherent *uncertainty* in the information, the model structure in which the information is presented, the assumptions bracketing the study, and any other constituent component require verbal descriptions with numerical detail where appropriate. As such, uncertainty statements can be both qualitative and quantitative.

Adhering to these ten data quality requirements is indispensable for LCA professionals especially when making public statements about the comparative ecological benefit of a product. Designers who critically evaluate the distinctions created by stated data quality assumptions in the goal and scope phase of a foundational LCA can better contextualize and evaluate the appropriateness of a published third-party-verified product declaration based on it.

5.3.11   What Are the Limitations of the Study?

Making design decisions based on an array of qualified facts, methods, and models laid out in the goal and scope phase may, at first seem, like a recipe for dithering. Appreciating the limitations on what any study can be used for, however, is essential to properly integrating them into a design workflow.

Broad statements regarding limitations contextualize the study for all stakeholders. Without them, users can draw inappropriate or even antithetical conclusions to what the study indicates. For example, a macro-level LCA based on economic input and output data cannot be used to tailor specific manufacturing processes since the data don't come from a process-based gathering methods and are only accurate when applied to evaluating impacts from a single general economic sector. A statement in the goal and scope description of a study, limiting its use to evaluating environmental impacts at the larger sectorial scale, safeguards it from being indiscriminately applied to the more fine-grained considerations needed to evaluate, modify, and improve a single industrial process.

Beyond any individual specific limitation stated in each study, it is critical to always keep in view LCA's numerous general methodological and data resource limitations. These will be discussed further in Chap. 6.

5.3.12   What Subjective Value Choices and Optional Elements
Beyond the Primary Inputs and Outputs Influence
the Study?

One general limitation in LCA practice comes from subjective value choices that individual practitioners make. It may surprise some designers, who regularly make subjective aesthetic and performative value judgments for a living and who are looking at LCA to guide them in more objective decision-making methods, to learn that subjectivity is a typical LCA consideration. The main difference between design and LCA may be how subjective value choices are explicitly stated and managed in a scientific framework. Where designers are not typically required to overtly identify and document what parts of their work is subjective and objective, best practices dictate that LCA practitioners do this.

Personal beliefs, attitudes, preferences, and the perception of risks determine a person's general perspective or outlook on the world. Perspective influences the value choices a practitioner makes about many fundamental elements in an LCA. Decisions regarding the type, amount, and quality of data to use, what characterization factors will be used to compare them, and what environmental impact categories are included or excluded all affect the results of the study.

An Maria De Schryver shows how the LCA community has used evolving cultural theory models to view and partially control for bias created by a range of personally held values. Individual perspectives regarding the natural world's general stability influencing choices can fall into five general architypes: the individualist, the hierarchist, the egalitarian, the fatalist, and the hermit. Of these, the hermit chooses to disengage from the world completely, and the fatalist perceives himself to be a victim to forces over which he has no agency and so neither is engaged in environmental decision-making. The individualist sees the world as supremely abundant and resilient. From this perspective, future consequences are negligible for any market-driven action that maximizes present value. The hierarchist recognizes the need to take responsibility for our collective actions in order to maintain nature's healthy equilibrium. Here, management decisions balance present and future value. The egalitarian sees nature as fragile and unstable and seeks solutions that, above all, do no harm to present and future generations.

De Schryver points out the limitations of these models and further cautions that practitioners, decision-makers, and other stakeholders do not usually neatly identify with a single archetype. Any of us can simultaneously straddle two or more of them or change our perspectives depending on our role in society at that time. Depending on the persistence of specific substances in the environment and impact categories considered, the practical difference between an individualist and an egalitarian perspective can vary widely and can substantially increase uncertainty in an impact assessment (De Schryver 2010).

While more research is underway in quantifying the effects of the wide range of value judgements, there are currently no natural science-based procedures that can be used to meaningfully aggregate them across impact categories. Fundamental qualitative statements regarding value choice can, however, be included and made explicit in the goal and scope phase when specifying the "obligatory product properties" of functional unit which directly attach key stakeholder preferences to the thing being studied (Weidema 2019).

5.3.13 How Will the Results of the Various Parts of the Study Be Systematically Interpreted?

Interpretation is the fourth phase listed in the LCA framework and can be understood as a bracketing "counterpart of the scientifically similar 'soft' phase 'definition of goal and scope' first phase" (Klöpffer and Grahl 2014). Knowing upfront what interpretive framework will be later applied to review each of the first three phases helps to set appropriate data sources and their quantity and quality requirements used in the LCI phase. It also helps when selecting which appropriate methods and specific evaluative LCIA phase components to use. Impact categories, category indicators, and characterization factors, for instance, will change depending on the end purpose of the study laid out in the goal and scope phase, and data coverage and quality standards increase or decrease as do the number of environmental impact categories included. ISO 14040 describes the interpretation phase of LCA to be that:

> …in which the findings from the inventory analysis and the impact assessment are considered together or, in the case of LCI studies, the findings of the inventory analysis only. The interpretation phase should deliver results that are consistent with the defined goal and scope and which reach conclusions, explain limitations and provide recommendations. (ISO 2006a)

The intentions of the interpretation phase first align goal and scope with the broad phase components included in or excluded from the study. In explaining the results of the study, they will largely reflect and reiterate the initial outline of the goal and scope study phase. Qualitative consistency between the first and fourth phase bracket and contextualize the conclusions produced by the quantitative second and third phase.

The necessary application of human judgment introduced early and late in a study can be seen by some as a fundamental weakness as compared to the methodological objective rigor at the core of any LCA (stages 2 and 3) or LCI (stage 2 only.) The judgements, however, are also informed and guided by other scientifically rigorous methods that serve as a backstop to drawing capricious or casual conclusions. Walter Klopffer sites Heijungs et al.'s five separate numerical approaches that can be applied in the interpretation phase. They include:

- Comparative analysis
- Contribution analysis or sectorial analysis
- Discernibility analysis
- Perturbation analysis
- Uncertainty analysis

Each analysis draws on different statistical methods to quantitatively evaluate and qualify the soundness of an interpretation (Heijungs et al. 2005).

It is rare for designers to run formal iterative LCA exercises in parallel to the primarily visual design process. Heuristic approaches, traditionally favored by the design profession, place a majority of trust in experience rather than on detailed analysis of discreet elements. Designers are not typically required to check and provide support for each decision made throughout their iterative workflow using statistical methods. When intentionally integrated into each visual iteration of the workflow, however, the product of the four indispensable phases in a full and proper LCA deepens the awareness of the environmental impacts of each design decision. Assessing the impacts that each material specification or required manufacturing process will have on the broader ecosystem is impossible without a formal protocol to structure the assessment and remains technically daunting even with one. The lack of consistent and reliable data covering all aspects of a finished product or system is only one component that increases uncertainty in an assessment. Stating this upfront is better than ignoring all information (imperfect) that would provide insight (maybe only marginally improved) to charting a path forward through a design project.

5.3.14   What Type of Final Report Is Required for the Study and How Is It to Be Formatted?

LCA professionals typically communicate results from formal LCA studies to a target audience through written reports. The structure of the report logically follows the structure of the study and covers all four phases of an LCA: goal and scope, life cycle inventory, life cycle impact assessment, and interpretation. In addition to an Executive Summary at the start of the report, four other sections are also typically included: Introduction, Methodology, Conclusions, and References. The structure of a report provides a clear outline describing all the key features and approaches used in the study. Beyond giving insight into the structure of the study, it is also useful to assess the similarities and differences between it and another or several other LCAs.

For LCA reports intended for *internal* consumption by the sponsor of the study, ISO 14044 does not specify required content except that:

> The results and conclusions… shall be completely and accurately reported without bias…The results, data, methods, assumptions and limitations shall be transparent and presented in sufficient detail to allow the reader to comprehend the complexities and trade-offs inherent in the LCA.

Final LCA reports take numerous forms and can include more or fewer details and qualifications depending on the final intents and purposes. An LCA intended for wider publication to anyone other than the study's commissioner or one that makes comparative assertions must meet a higher standard of care in reporting the findings. Flanagan and Ingwersen provide an ideal Report Table of Contents:

Executive Summary
1. Introduction
    1.1. Study background and context
    1.2. Introduction to LCA
2. Goal and Scope
    2.1. Goal
    2.2. Function, functional unit, and reference flow
    2.3. System boundaries
    2.3.1. General system description
    2.3.2. Description of unit processes (qualitative)
    2.3.3. System boundaries
    2.4. Cut-off criteria
3. Methodology
    3.1. Assumptions
    3.2. Limitations
    3.2.1. Limitations of LCA methodology
    3.2.2. Limitations of the study
    3.3. Life Cycle Inventory
    3.3.1. Data collection procedures
    3.3.2. Description of unit processes (quantitative)
    3.3.3. Data sources
    3.3.4. Data quality requirements
    3.3.5. Treatment of missing data
    3.3.6. Calculation procedures
    3.3.7. Allocation
    3.3.8. Geographic and temporal relevance
    3.4. Life Cycle Impact Assessment (LCIA)
    3.4.1. Life Cycle Impact Assessment methods
    3.4.2. Impact categories and category indicators
    3.4.3. Normalization (if used)
    3.4.4. Grouping (if used)
    3.4.5. Weighting (if used)
    3.5. Calculation Tool
    3.6. Critical Review (if performed)
4. Life Cycle Inventory (LCI) results
5. Life Cycle Impact Assessment (LCIA) results
    5.1. Comparative analysis per functional unit
    5.2. Contribution analysis
6. Interpretation
    6.1.Key findings
    6.2.Evaluation
    6.2.1. Completeness Check
    6.2.2. Consistency Check
    6.2.3. Data Quality Analysis
    6.2.4. Sensitivity Analysis
    6.2.5. Uncertainty Analysis
7. Conclusions and recommendations
8. References
Appendices
    LCI data tables
    LCIA characterization tables

> *Critical Review*
> *Names and affiliations of reviewers*
> *Critical review reports*
> *Responses to critical review recommendations*
> *Final statement from Critical Review panel chair*
> (Flanagan and Ingwersen 2014)

Sections 1 through 3 of this illustrative Table of Contents comprise the first, primarily qualitative, part of the report and introduce Sections 4 and 5 which contain most of the report's quantitative information. Sections 6 and 7 conclude with an interpretation and recommendations based on the study's contents. Understanding the structure of these reports reduces the time required by any third party including a design professional to review, compare, and use their findings. Details about individual sections and the formats typically used to graphically communicate findings are included in Chaps. 8 and 9 of this book.

5.3.15 Is a Third-Party Critical Review Required to Validate the Final Report?

The first and second parties in any LCA are, respectively, the commissioner of the report and the person or agency who prepares it. These two parties must work closely together when primary data about proprietary or closely held industrial production processes are used to offer insight about their environmental impacts or possible improvements. In these cases, nondisclosure agreements are used to protect trade secrets, and the final report remains in the hands of the first party as an internal document. This was the case 50 years ago for the very first commercial, proto-LCA, REPA commissioned by the Coca-Cola Company in 1969 whose details have never been shared with the public.

Third parties include those whose job is to critically review the final report for methodological consistency with ISO standards and appropriateness. They also include readers with an interest in its contents and conclusions. LCA reports that make public comparative assertions that argue for one product over another must include a section on the critical review process and include a list of reviewer names and affiliations (ISO 14044 2006b). Designers are third-party LCA consumers when reading documents such as environmental product declarations (EPDs) or using software built on formal LCA activities. The ability to identify and differentiate between higher- and lower-quality LCA information rests largely on knowing what is in a complete report verified by another expert third party.

Many companies and individuals provide credible expert third-party review. Even the firm that ran the 1969 study for Coca-Cola, Franklin Associates, still offers third-party verification services but is now only one small player in a burgeoning field of global consultancies. In all cases, interpreting statements and assertions regarding an individual product's environmental impact requires the ability to trace back its underlying facts to a trustworthy source within a proper organizing framework.

References

An Maria De Schryver (2010) Value choices in life cycle impact assessment. Phd thesis, Radboud University, Nijmegen

Curran MA (ed) (2012) Life cycle assessment handbook: a guide for environmentally sustainable products, 1st edn. Wiley-Scrivener, Hoboken

Frischknecht R (1998) Life cycle inventory analysis for decision-making: scope-dependent inventory system models and context-specific joint product allocation. ETH

Hauschild MZ, Rosenbaum RK, Olsen SI (2018) Life cycle assessment: theory and practice. Springer, Cham

Heijungs R, Suh S, Kleijn R (2005) Numerical approaches to life cycle interpretation - the case of the Ecoinvent'96 database (10 pp). Int J Life Cycle Assess 10:103–112. https://doi.org/10.1065/lca2004.06.161

Ingwersen W, Flanagan W (2014) Communicating LCA results. In: Shenck R, White P (eds) Environmental life cycle assessment: measuring the environmental performance of products. American Center for Life Cycle Assessment (ACLCA), Vashon, p 236

ISO (2006a) ISO 14040 environmental management: life cycle assessment: principles and framework. International Organization for Standardization, Geneva

ISO (2006b) ISO 14044 environmental management: life cycle assessment: requirements and guidelines. International Organization for Standardization, Geneva

Jolliet O, Saade-Sbeih M, Shaked S et al (2015) Environmental life cycle assessment, 1st edn. CRC Press, Boca Raton

Klöpffer W (2006) The role of SETAC in the development of LCA. Int J Life Cycle Assess 11:116–122. https://doi.org/10.1065/lca2006.04.019

Klöpffer W, Grahl B (2014) Life Cycle Assessment (LCA): a guide to best practice, 1st edn. Wiley-VCH, Weinheim

Masson-Delmotte V, Zhai P, Pörtner D, et al (2018) Summary for policymakers. In: Global warming of 1.5°C. An IPCC special report on the impacts of global warming of 1.5°C above pre-industrial levels and related global greenhouse gas emission pathways, in the context of strengthening the global response to the threat of climate change, sustainable development, and efforts to eradicate poverty

Matthews HS, Hendrickson CT, Mathews D (2014) Life cycle assessment: quantitative approaches for decisions that matter. Open access textbook

NJDEP-Air Quality Management (2018) https://www.state.nj.us/dep/aqm/es/emstatpg.html. Accessed 18 Jul 2019

Rebitzer G, Ekvall T, Frischknecht R et al (2004) Life cycle assessment: part 1: framework, goal and scope definition, inventory analysis, and applications. Environ Int 30:701–720. https://doi.org/10.1016/j.envint.2003.11.005

Rodriguez BX, Simonen K, Huang M, De Wolf C (2019) A taxonomy for Whole Building Life Cycle Assessment (WBLCA). Smart Sustain Built Environ 8:190–205. https://doi.org/10.1108/SASBE-06-2018-0034

SETAC (1994) Guidelines for Life-Cycle Assessment. Environ Sci Pollut Res 1:55–55. https://doi.org/10.1007/BF02986927

Simonen K (2014) Life cycle assessment, 1st edn. Routledge, London/New York

Sullivan L (1896) The Tall Office building artistically reconsidered. Lippincott's Magazine 403–409 US EPA O (2016) Understanding global warming potentials. In: US EPA. https://www.epa.gov/ghgemissions/understanding-global-warming-potentials. Accessed 27 Sep 2020

US EPA (2016) Understanding global warming potentials. In: US EPA. https://www.epa.gov/ghgemissions/understanding-global-warming-potentials

van Oers L, de Koning A, Guinée J, Huppes G (2002) Abiotic resource depletion in LCA: Improving characterisation factors for abiotic resource depletion as recommended in the new Dutch LCA Handbook

Weidema BP (2019) Consistency check for life cycle assessments. Int J Life Cycle Assess 24:926–934. https://doi.org/10.1007/s11367-018-1542-9

Chapter 6
Addressing Resistance to a Fact-Based Approach

Abstract Facts are critical to making informed decisions but are often not enough to convince people to change their priorities. Scientific approaches like LCA presents obstacles and opportunities to non-specialists, including designers.

The self-preservation instinct is typically hardwired into each individual. This genetic predisposition to avoid certain risks that loom large in our consciousness both safeguards the continuation of our species and blinds us to less visible and possibly much greater threats. Media outlets play to the natural human reliance on heuristics and implicit biases. The confusion a "post-fact" world creates serves to maintain the status quo by perpetually calling into question well-established facts and stalling fruitful debate. Spectacular headlines provoke irrational fears that distract from other, more statistically likely threats.

Rhetorical techniques frame facts. Denialism is intended to create paralysis. More information is not necessarily better, if it is derived from sources that only confirm sociopolitical biases rather than challenging them. It is important to acknowledge contextual headwinds operating at the largest scales in society before blithely applying a targeted set of analytical techniques to measure environmental impacts in a design project.

Before presenting the "how-to" of integrating a data-driven approach to sustainable design decisions, however, it is important to recognize that facts in one domain are not always enough to persuade a decision-maker. Attributional LCA may improve process-based decisions without requiring wholesale recanting of firmly held world views. Non-geometric data serves designers as a foundation to a fact-based and measured approach.

6.1 Fast and Slow Decisions

Every day, we decide on whether to buy this or that, to drive faster or slower, or to vote for one candidate over another. We make certain decisions instantaneously, like running from a snarling dog, stepping out of the way of a speeding bus, or picking a favorite toothpaste from a grocery shelf full of competing products. These decisions leverage what psychologists refer to System 1 thinking. We reach them quickly and automatically, and they require little effort or concentration. Other decisions or activities require careful evaluation and calculation. Deliberating over which

© Springer Nature Switzerland AG 2021
J. Cays, *An Environmental Life Cycle Approach to Design*,
https://doi.org/10.1007/978-3-030-63802-3_6

healthcare plan to enroll our family in, for example, requires System 2 thinking to take over. System 2 engages when reaching a conclusion requires analytical processes (Kahneman 2013). Material decisions based in patient and technically detailed life cycle assessment activities reside primarily in the System 2 domain.

Sometimes the perceived importance of the decision and the weight of making the correct decision are so great that we freeze. We fall victim to the "paralysis of analysis." Out of the fear of making the wrong choice, we get stuck in an endless loop of comparing the relative costs and benefits of each option. As the number of options rises, or the decision-making process identifies conflicting criteria, the time spent trying to decide also increases.

At some point we might conclude that making no decision is a choice and either walk away from the weighty responsibility or arbitrarily pick one option from an array of choices. Complicated choices, however, do not make up the majority of daily decision-making. Most of the choices we make every day seem relatively trivial, but taken together, the effects can add up. What we feel is risky behavior can have a lot to do with how we perceive it will impact our lives and, perhaps most importantly, when the effects will manifest.

6.1.1 Sustainable Choices

Behavioral economists study the way people make decisions. By observing people's habits and the outcomes of their decisions in multiple settings, they collect and interpret data in order to more accurately predict the odds that most people will make one choice over another. The insights they uncover about our natural inclinations are based on observations of subjects in controlled experiments or pointed surveys that measure not the way we ought to behave but how we actually do behave when given the chance to decide between two or more options. Studies have shown that what we want is dependent on how close to now we are going to get it. We tend to make choices that conform to what we have been told is good for us. The healthy choice, the one that promises to bring about a state of well-being and long-term benefits, is what we tend to choose when we are planning for the future. The farther off "the future" is, the more likely we are to choose the rational, the healthy, and the "good" options, when focused on that which directly affects our own well-being. Our natural inclinations are focused primarily on what is closest to us in time and space. Anything that might indirectly affect us is on the perceptual back burner.

Harvard behavioral economist, David Laibson, speaks about "intertemporal choice" and how our actions or purchasing decisions can positively or negatively affect us in the future:

> There's a fundamental tension, in humans and other animals, between seizing available rewards in the present, and being patient for rewards in the future," he says. "It's radically important. People very robustly want instant gratification right now, and want to be patient in the future. If you ask people, 'Which do you want right now, fruit or chocolate?' they say, 'Chocolate!' But if you ask, 'Which one a week from now?' they will say, 'Fruit.' Now we

want chocolate, cigarettes, and a trashy movie. In the future, we want to eat fruit, to quit smoking, and to watch Bergman films. (Lambert 2006)

This decision-making dynamic holds in the considerations that dominate the commercial design world as well, where short-term risks of financial failure outweigh long-term environmental or social concerns. Clients responsible for the financial performance of individual projects and initiatives may make aspirational statements that embrace a sustainable design ethos, but meaningful actions are not likely to follow if they require any real change in business behavior or priorities.

Heuristics are used to substitute complex questions with simpler ones to break the cycle of decision-making paralysis and arrive at an answer. Designers may reframe an inherently complex problem using only a subset of criteria they feel most comfortable with to define acceptable parameters. When faced with complexity, or a threat to the financial bottom line, it is common to retreat to the traditional metrics of success.

Sustainable design and production is a multifaceted domain that ideally would include triple-bottom-line accounting considerations (Elkington 1998). To remain in business, however, mainstream design practices continue to hew closely to a client's primary concerns which are often financial. A design firm that consistently and completely ignores financial, single-bottom-line issues is not viable in the marketplace for long.

Even designers able to successfully expand a single-bottom-line focus to include a project's environmental impact often simplify the issues under consideration. To avoid the complexity that a multivariable matrix of weighted and qualified environmental impact categories introduces, they may optimize a single performance criterion such as energy efficiency, carbon foot print, or waste reduction. Attempting to treat all impact categories equally would require time and effort that are currently too resource intensive to integrate into the design process. The next three chapters, however, present workflow options to realistically consider a wider subset of material environmental impacts using purpose-built tools to do so.

6.1.2 Biases

Cognitive biases are constantly at work in the evaluation of relative threats or benefits. Availability biases work to convince that an event is less likely to occur if there are no examples available in recent memory; tomorrow's world will be a lot like yesterday's world based on easily retrieved memories (Schwartz et al. 1991). Confirmation biases promote the searching and accepting of only that information that supports personal view, beliefs, or values (Nickerson 1998). Hyperbolic discounting is at work in the earlier "fruit or chocolate" example when making more virtuous or health choices later but take immediate payoffs now. People make choices today that their future selves would prefer not to have made, despite using the same reasoning (Laibson 1997). Normalcy biases lead people to minimize or completely disbelieve threat warnings. "The initial response to a disaster warning is disbelief" (Drabek 1986). Biases are the lenses we all filter information through when making decisions, and they warp perception when assessing likely threats and opportunities.

In his book, *Global Catastrophes and Trends, the Next 50 Years*, Vaclav Smil presents a measured and thoroughly researched evaluation of an array of possible mid-twenty-first-century threats. His approach systematically considers the odds that various natural and cultural/technological risks have to negatively affect the lives of individual human beings or wipe out the entire human civilization. Smil shows, for example, how relatively unlikely people were to die from a volcanic eruption, airline accident, or terrorist attack compared to car accidents or medical errors which were between three and four orders of magnitude more likely to take a life (Smil 2008). People tend to not focus on and guard against less spectacular but more likely risk events.[1]

6.2   Climate Change Story

Directly explaining climate change to the public is like telling a story about rust. The pace and technical details of the story make capturing the imagination a challenge. No matter how certain the science is about what drives it, the facts and phenomena play out over a time scale to which most people have trouble relating or caring about. Even the most committed designers will likely not change the hearts and minds of clients who are not already convinced that there is a problem that their design addresses.

Stories of calamity and destruction are best told when the action happens suddenly. Unlike a nuclear explosion, a plane crash, or a house fire, the story about the incremental changes in the ratio of certain harmful chemical elements and compounds in air and water, measured in parts per billion, does not immediately captivate. Even when it concerns all the air and water we have, the slow pace and amorphous, invisible action tend to lose people's attention. In today's 24-hour news cycle, large and spectacular short-term changes that directly affect human life for the worse make better stories. They are fixed in the imagination more effectively than long, drawn out, "boring" processes that indirectly cause harm. The steadily warming environmental conditions that contribute to storm intensity (Murakami et al. 2017) don't generate the same interest or audience ratings as the sensational coverage of a hurricane (Grabe et al. 2001). Because the violent effects of such a storm are clearly tragic, they focus attention and immediate action to respond to the crisis.

---

[1] In the 2008 book, Smil concludes that we should be preparing for a nearly 100% chance of a novel strain of influenza to produce a global pandemic. He also calls for government programs to simultaneously increase energy efficiency and decrease overall use to benefit the environment.

Energy use, as discussed in Chap. 2, underpins every climate change story. Generating and using increasing amounts of the energy required to maintain the consumption/production status quo is directly linked to a network of complex environmental issues slowly playing out in the ecosphere (US EPA 2016). Attachment to habits are the root cause of increased energy consumption. In order to avoid modifying behavior born out of ideas of what we want, need, and deserve, now and in the future, some engage in endless debate to stall action (Oreskes and Conway 2011). While we wait, mean global temperatures continue to rise.

For those who wish to stall, what is said in these opposing narratives is not as important as how it is said in order to keep the debate going. If there are a hundred steps needed to effectively address the actual crisis, the debate functions to keep a long pause on the first step. While everyone is frozen, arguing the fundamentals, concerted large-scale action is impossible. Democratically elected governments wait, hedge bets for political and short-term economic reasons, and stall. Agencies, tasked with studying the problem further, make incremental policy changes and write regulations that may or may not be enforceable. Effective action to address this problem is hindered, while the fundamental cause of the problem is endlessly debated. Despite the majority of international scientists reporting overwhelming evidence to support the view that human activity is causing global temperature to rise, the debate continues.

Although gradual and incremental, sea-level rise may be the most visible of all the effects, as higher seas directly impact coastal cities. Buildings and entire regions, flooded or washed away by ever-costlier storms, make headlines. Rather than spurring action to address root causes, threats from large storm surges mobilize governments who have the means to "harden" infrastructure with steel and concrete barriers against hurricane force winds and waves. Wealthy nations fund resiliency initiatives to protect life and property in vulnerable areas where they have economic or political interest all over the world.

Other effects, while even less evident, are no less real and even more dangerous to life on earth. Subtle but irreversible changes to the chemical composition of the same seawater that inundates shore communities have farther-reaching implications. These subtle changes extend beyond the shore to the air we breathe and the soil we depend on to feed ourselves. They affect everyone, not only coastal dwellers. No matter how high up or how far inland we move from flood-prone cities, we still need clean air, potable water, and healthy food to live.

6.2.1   The False Debate over Fundamental Causes of Climate Change

Some people, with the potential to influence policy decisions, do not or will not believe human activity has any significant impact on the overall environment. Some may acknowledge the connection between the two but will not act. Many of the same lawmakers who pass funding initiatives to build higher seawalls against rising sea levels, and even agree to fortify natural estuary buffers to address a threat whose effects they can see, lack the vision or political will to tackle the root causes of the threat (Nawaguna 2020). They may not *fully* subscribe to a maximum fossil fuel extraction agenda powering established production and consumption patterns. Nevertheless, the economic imperative to maintain and support existing carbon-based energy production systems are often too politically expedient and compelling to argue against publically. The net effect is that they do not publicly accept climate change as a dilemma. They remain unconcerned with the consequences of increasing the extraction and burning of fossil fuels, so long as the economy in their district grows enough to get them reelected. This attitude forms one pole of the debate, and it elicits an equally strong opposite reaction from those who disagree (Tyson and Kennedy 2020).

Still, most people, irrespective of which opinion they hold, simply do not know for sure. From time to time, they may tune in to the climate change debate to see who is prevailing. Ideally, the debate would be rooted in hard science and the consensus findings of the great majority of researchers. The enormous short-term political and economic outcomes at stake, however, often trump rational serious consideration of scientific facts. Specious arguments and denialism attempt to obfuscate decades of sound and patient global climate research founded on centuries of fundamental and uncontroversial science.

These two opposing views are not represented equally among people in the scientific community. An overwhelming majority of the world's climate research scientists (97%) have concluded that the rise in global temperatures and the resultant change in our climate systems are caused by human activity (Cook et al. 2013; Cook et al. 2016; NASA 2020). Still, denialist fringe groups persist in formulating messages intended to maximize doubt in debates carried by both conservative and liberal media outlets. Such false skepticism, promoted to create the net perception that we still do not know for sure what is causing the rise in global temperatures, sidesteps the scientific debate. Climate denialists regularly use pseudoscience to directly confront genuine climate science (Hannson 2017).

Conservative media backs status-quo industry interests and regularly parades unqualified "experts" to opine on the debate and cast doubt on the scientific consensus in order to keep advertising funding flowing. Some of their chosen spokespeople have science credentials (not necessarily climate science), but many are chosen primarily for their considerable rhetorical, debate, and presentation skills. In the name of fair and balanced coverage, even mainstream media outlets may turn to the same people when reporting on climate change issues.

Denialists use a set of common rhetorical techniques in the media that leverage normal human biases to challenge scientific facts. John Cook is a research assistant professor at George Mason University's Center for Climate Change Communication. He has developed the mnemonic acronym, FLICC, to help people clearly remember which ones are being used to frame arguments to refute scientific facts. His FLICC taxonomy uses a series of subtitled icons to show and reinforce the array of techniques used (see Fig. 6.1)

Techniques of Science Denial

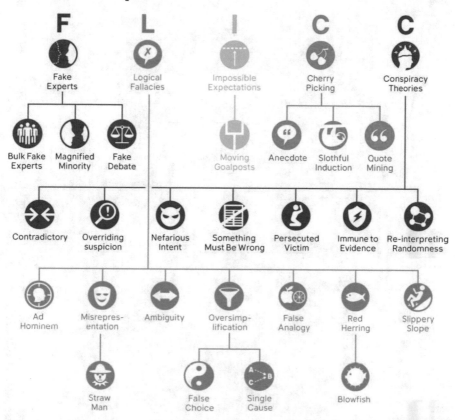

Fig. 6.1 FLICC taxonomy of techniques found in science denial showing the five main techniques used to deny science. They can be remembered with the acronym FLICC: fake experts, logical fallacies, impossible expectations, cherry picking, conspiracy theories. Each has one or more sub-fallacies shown in the taxonomy. (Used under Creative Commons Attribution-Share Alike 4.0 International)

Instead of proceeding to fruitful discussions about what specific corrective actions to take to improve our ecological fortunes, we allow ourselves to become mired in conversations about fundamental science. The physical causes of a rise in atmospheric temperatures have been known for over a century, ever since Svante Arrhenius published his quantification of carbon dioxide's contribution to the greenhouse effect (Arrhenius 1896).

The reason the debate can continue at all, though, is that most people do not know anything about fundamental climate science. This is broadly true of anyone outside the scientific community and includes policymakers, corporate leaders, consumers, and designers. Figure 6.2 clearly shows the climate science milestone

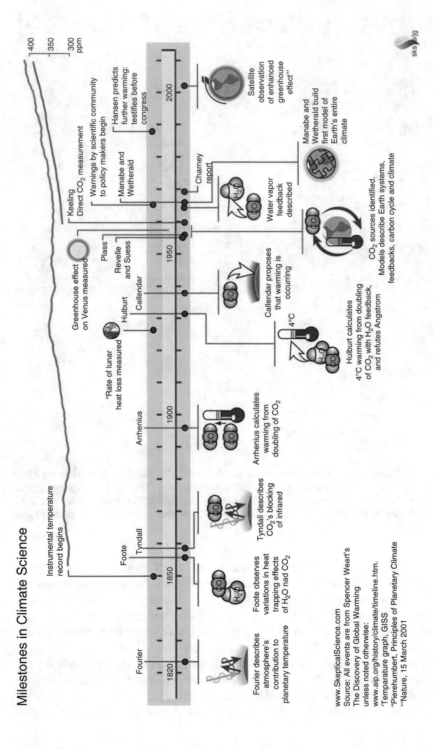

Fig. 6.2 Timeline of major events in the history of climate science, 1820 to present. From "The History of Climate Science" on Skeptical Science: http://sks.to/history (Cook et al. 2016). (Used under Creative Commons Attribution-Share Alike 4.0 International)

discoveries that for nearly 200 years have established clear evidence linking increasing CO_2 levels to global warming.

The protracted debates, however, do not continue due to a lack of information or knowledge. It may not even come down to a denial that a crisis exists. Rather, ecological concerns may simply be significantly lower priorities than economic ones. Hyperbolic discounting, operating at industrial scale, needs to be acknowledged and addressed. If both ecologically and economically acceptable product and project options are available, then the need for a tradeoff between them disappears. This is applicable not only to design but to all industrial disciplines. Environmental LCA can play a part by providing midpoint indicators of environmental performance. But there are further considerations to deal with before committing to building LCA into a design workflow.

6.3   Taking Issue with Life Cycle Assessment

Many critics have pointed to the basic problem of data reliability and the effort it takes to properly collect and use it as LCA's Achilles heel. In her overview treatment of the subject as it relates to building design and building products, Kathrina Simonen correctly identifies barriers to fundamental accuracy, usefulness, and cost-effectiveness the results issuing from study documents. She points out the arguable pessimists' view that LCA is being oversold and that broad extrapolations can be consistently drawn and used from existing expensive and detailed LCA studies:

> A huge effort is required to conduct an LCA and the results can be of questionable value: uncertain with limited value to conduct change. One can argue that even in the relatively certain field of operational energy use (with meters to track consumption), we find our ability to predict actual energy use has limited precision. How can we expect to develop credible data throughout a complex and changing supply chain and the uncertainty of building life cycle stages?...Unless LCA can become *cost-effective* and integrated within existing design and manufacturing processes, the effort of conducting an LCA can often outweigh the benefits. (Simonen 2014)

This pointed criticism regarding the inherent complexity, uncertainty, and great expense in time and money to conduct a full life cycle assessment is shared across disciplines and will most likely not be solved by any single industry, profession, or field. LCA conclusions also have the disadvantage of being applicable or valid only to the specific product or service it studied. Change one materially important process or boundary condition in the defined life cycle and the whole profile can change, rendering the formal LCA no longer valid to describe what it originally assessed.

The first iterative LCA "goal and scope" phase anticipates and foreshadows the next three. Before proceeding with a discussion and demonstration of beneficial applications of subsequent LCA phases, it may help to pause and address the questions and concerns raised by this overall approach to evaluating what and how to satisfy the demands of a market. Since the majority of both designers and clients are not steeped in the technical ins and outs of ecologically responsible evaluations, some may hesitate to embrace an additional process that seems to demand:

1. Extra expense.
2. A faith in a set of detailed and qualified facts and study conclusions expressed in scientific terms.
3. Willingness to expand focus and attention beyond a single ecologically important impact category such as global warming.
4. A willingness at the end of it all to accept a level of uncertainty that decisions will actually yield ecologically superior decisions across multiple impact categories.
5. Acceptance that LCA focusses primarily on environmental impacts and does not take other important considerations such as cost and societal impacts into consideration.
6. Cannot effect an order of magnitude change through available options.
7. That there may be no truly "good" decisions and "less bad" ones will only delay the inevitable.
8. Acceptance of results that are not comparable with any other study document of a similar product system.
9. Faith in imperfect models that are abstract and that cannot fully capture reality.
10. Comfort with nonintuitive, nonvisual, data-driven processes to provide insight.

These are all legitimate concerns that, if ignored, will hamper efforts to expand the use of LCA and the design tools that it supports. New tools used to integrate LCA data into the design process can mitigate some of the concerns but until there is a fully automated and seamless interface, challenges to broad acceptance will continue to persist. Yale University's Phillip Bernstein predicts that as workflows shift to more cloud-based integrated data-centric models with a constellation of tools accessing a common data feed, a greater number of firms will be able to directly integrate LCA into architectural projects (Cays 2017).

6.3.1   The Attributional LCA Time Scale

Life cycle assessment provides foundational information leading to better predictions within a clearly defined time horizon and potential impacts within that temporal window. While nearly impossible to directly compare and evaluate all aggregate risks associated with industrial product environmental impacts, it is possible to at least see and compare two or more options in a study before deciding on which of them to choose. Two different but legitimate approaches are available but serve different purposes. The ILCD handbook states that:

> The attributional LCI model describes its actual or forecasted specific or average supply chain plus its use and end-of-life value chain, all embedded into a static technosphere.

> The consequential LCI model describes the supply chain as it is theoretically expected in consequence of the analyzed decision, embedded in a dynamic technosphere that reacts to a change in the demand for different products (EUR 24708 EN 2010).

Potential impacts to midpoint impact categories are calculated from life cycle inventory data. Midpoint categories include:

Climate change.
Ozone depletion.
Human toxicity.
Respiratory inorganics.
Ionizing radiation.
Ground-level photochemical ozone formation.
Acidification.
Eutrophication.
Ecotoxicity.
Land use.
Resource depletion.

These midpoint categories can then be extended further out on the impact pathway to broad endpoint categories such as damage to human health, damage to ecosystem diversity, and resource scarcity. The farther into the future, the wider the spatial boundaries, and the greater the quantity of interacting variables in the system being studied, the less certain any attempt will be to accurately calculate the probability of any specific consequence coming to pass. It is likely that consequential LCA will continue to be challenged by technical limits, complexity, and inherent uncertainty that surround making large-scale predictions. For designers, the tools available primarily use attributional methodologies that focus on potential midpoint impacts inferred from LCI data.

6.4 An Appeal to Reason

When doubt and confusion reign, decision-making suffers. As we struggle to define and delineate the boundaries of such a large and complex problem as climate change, there can be an enormous delay before we even start developing effective approaches to its solution. Before we decide on even the first steps, the problem itself must be firmly established and understood. Questioning the causes is where most of us get stuck, paralyzed by the conflicting messages we hear from the "experts." Neither of the dominant opposing views refutes the fact that the average temperature on the planet is rising. However, global warming is alternately referred to as a human-caused "crisis" demanding immediate action or as part of a natural process over which we have absolutely no control.

Effective appeals to consider alternatives to operational status quo often fail to convince. Emotionally charged positions based solely on firmly held personal beliefs rarely change when confronted with facts (Björnberg et al. 2017). This is especially true when there is a perceived potential for immediate or short-term financial loss. There is no better illustration of this dysfunctional dynamic between ideology and facts than the variety of politically motivated COVID-19 global pandemic responses and the results playing out in 2019, 2020, and probably well into 2021 even as life-saving vaccinations are made widely available.

World views, immune to evidence-based arguments, are built over a lifetime shaped by experience that supports counterfactual reasoning (Van Hoeck et al. 2015). While fact-based approaches offer limited success in prompting a change in the behavior of the most strident climate and environmental science denialists, they, nevertheless, provide a foundation to support rational action. The LCA framework provides a stable structure to present detailed evidence of the environmental effects attached to any product system. Framed by the formally established goal and scope phase, subsequent LCA phases provide more detail and reach qualified conclusions about the impacts of the assessed goods and services.

The parallels between LCA and design praxis in establishing a well-defined project goal and scope don't hold for the critical LCI and LCIA phases. Designers presenting evidence-based sustainable options to clients are not typically involved in these detailed technical phases but, as discussed in the next three chapters, can employ tools and techniques supported by the mounting evidence and growing dataset continuously generated by experienced LCA practitioners and the industries they study closely.

Deliberative thought, predicated on trustworthy facts, requires time. The practices and language we have developed for thinking, reasoning, and collectively deciding are methodical and thorough. In a post-fact media landscape, however, we are encouraged to do nothing. But, as established in Chap. 1, that is an impossibility. While we allow business to go on as usual, each living breathing human being on the planet continues to take in and give back, and our technologically amplified actions are increasingly unsustainable.

6.4.1   Impossible to Do Nothing

With that in mind, it is worth noting that there are two meanings to the phrase "impossible to do nothing." The first is a biological fact of being alive here and now. The second is a call to act, to first understand the scope and consequences of our thoughts and actions, and then to make the necessary and appropriate changes that will ensure our own and future generations' ability to thrive. Effective changes take place at different rates through individual daily decisions on what and how we will consume and also through collective political action.

It is also a fact that, whatever the causes, *inaction* at the individual and collective level is a decision with consequences. The status quo has inertia on its side to maintain deeply engrained habits and practices just as they are. Any change in direction to the system, even a small one, requires a force to move it. Therefore, before we design anything, we must first understand the key elements that define and influence the system we wish to improve; then we must improve it through decisive action informed by trustworthy data.

6.4.2 Inventory Analysis (LCI)

Data is at the heart of all LCA activities. LCA experts define life cycle inventory (LCI) as "a methodology for estimating the consumption of resources and the quantities of waste flows and emissions caused by or otherwise attributable to a product's life cycle. Consumption of resources and generation of waste/emissions are likely to occur at multiple sites and regions of the world, as different fractions of the total emissions at any one site (the fraction required to provide the specified functional unit; allocation amongst related and nonrelated co-products in a facility such as a refinery, etc.), at different times (e.g. use phase of a car compared to its disposal), and over different time periods (multiple generations in some cases, e.g. for landfilling)" (Rebitzer et al. 2004). ISO 14040 and ISO 14044 provide guidance in how to carry out LCI activities.

LCA practitioners have, since the 1990s, taken data quality into consideration. Established data pedigree matrices qualify individual bits of information that flow into the models that describe product systems. Pedigree typically includes indicators such as source reliability, completeness, temporal correlation, geographical correlation, and technological correlation (Weidema 1998; Frischknecht et al. 2005). Chap. 7 will further discuss the essential LCI data foundations and how they support design workflows tools.

6.5 Dominance of Geometric Data Models in Design Education

Even when there is general sympathy with integrating new and better data-driven approaches, established visual cultural practices may resist change. Just as the single financial bottom line can dominate all other concerns in professional design practices, the aesthetic bottom line typically reigns in schools of design where financial constraints can be disregarded.

There are both challenges and opportunities in the wider adoption of LCA in professional architectural education. In an Architectural Design article on LCA data and design workflows, I noted that "Life Cycle Assessment is fundamentally abstract and technical, as opposed to visual and intuitive, which may account for its slow adoption by most architecture firms" (Cays 2017). This observation continues to be reinforced by the conversations I have with designers and design students today.

Models of architecture and design projects in both the academy and in professional practice focus overwhelmingly on the geometric relationships of proposed design elements. Final presentation diagrams, technical drawings, and renderings in design critiques focus on the physical dimensions, programmatic arrangements, and functions supporting a design concept. Students learn early to focus primarily on the visual aspects of their projects as a matter of course.

The largely qualitative academic and professional design discussions contrast starkly with LCA's dry quantitative technical language and heavily qualified statements. Interpretive data visualization tools and dashboards make LCA, nongeometric, tabular data more accessible. However, the physically large-scale, but essentially invisible, environmental phenomena these data describe rarely turn heads in a design critique. Juries discuss the formal merits of a design project. Asking panelists to focus on the "eutrophication potential" or even the more recognized and accepted categorical threat "global warming potential" of different design alternatives is a hard sell. What to do with environmental impact assessment statements, even when based on trustworthy data, is not immediately evident.

6.5.1 Aesthetic/Technical Split in the Academy

Professional design schools teach important technical subjects, but, in the end, these are often relegated to relatively minor roles in holistic design assessment. In architectural education, building science courses in construction, structures, and environmental control systems supports the design studio where a two and a half to one lion's share of credits in design curricula are distributed. Design programs typically attract students who want to manifest their creative impulses at various scales in the physical world. Technical analysis is typically tolerated but less valued than formal synthesis. This is the main challenge to integrate LCA into design curricula.

The tension between the technical and aesthetic reflects the bifurcated origins of the design academy. Today's design schools are rooted in and modeled on two parallel academic traditions: the polytechnic and the beaux arts. In today's siloed academic environment, integrating the "Arts and Sciences" as a set of complementary approaches to explore physical phenomena is still more of a wish than a practical reality. This is true even in amalgamated and applied disciplines like design.

There is a growing student and faculty interest in "life cycle thinking." But when faced with the requirements of conducting proper studies as defined in the ISO 14040, 14,044, and other data-driven standards, the conversation quickly turns away from rigorous quantitative analysis back to the familiar comfort and safety of formal synthesis. This is, in no way, intended to diminish the necessity of serious qualitative design critique. Rather, it is an observation of primary drivers of academic design decisions that present a challenge for LCA adoption.

6.5.2   Bridging the Divide

In the 2012 Life Cycle Assessment Handbook's final chapter titled "Life Cycle Knowledge Informs Greener Products," James Fava identifies the emergence of software interfaces that will promote wider LCA adoption by the design disciplines and "go a long way toward ensuring that life cycle environmental information is used in product commercialization" (Fava 2012). This practical approach recognizes that it is unrealistic to expect non-specialists to embrace the rigors and statistical subtleties involved in formal LCA goal and scope definition, data inventory collection, impact assessment, and iterative interpretation exercises. Fortunately, software interfaces that marry digital geometric models with non-geometric data and provide impact assessments over multiple environmental impact categories are becoming increasingly available. Additionally, CPU and GPU hardware proceessing power continues its steady climb, setting the groundwork for a new generation of real-time interactive dashboards. These new interfaces will potentially be able to draw LCI data in and integrate it directly into the material libraries in designers' geometric models.

6.5.3   One Country's Call to Measure and Assess

Calls are growing louder by both the design professions and the academy to have sustainability data-drive design. The American Institute of Architects' 2019 Framework for Design Excellence and the National Architectural Accrediting Board's 2020 Conditions for Accreditation both promote actionable sustainability standards. The AIA Framework's Design for Resources provides useful links to LCA software resources and the International EPD Database. These voices are added to numerous other professional academic, professional, and third sector groups that have, for many years, been advocating for a science- and data-driven approach to a more sustainably designed built environment.

In the summer of 2019, over 100 representative academic, professional, regulator, and student stakeholders gathered in Chicago. During a 3-day multilateral forum. They finalized the main points driving the Conditions of Architectural Accreditation, and a writing team of seven condensed the stakeholder findings into the final 2020 Conditions for Accreditation. Six shared overarching values were defined. One of which is environmental stewardship and professional responsibility. It says that:

> Architects are responsible for the impact of their work on the natural world and on public health, safety, and welfare. As professionals and designers of the built environment, we embrace these responsibilities and act ethically to accomplish them.

A fine statement but in order to be actionable, two more criteria, to be met by all architecture programs, were included in the 2020 conditions:

1. PC.3 Ecological Knowledge and Responsibility which States that

> A program must demonstrate how its curriculum, structure, and other experiences address[es … how it] instills in students a holistic understanding of the dynamic between built and natural environments, enabling future architects to mitigate climate change responsibly by leveraging ecological, advanced building performance, adaptation, and resilience principles in their work and advocacy activities.

2. SC.5 Design Synthesis

> [A program must demonstrate how it ensures, through program curricula and other experiences, with an emphasis on the articulation of learning objectives and assessment] that students develop the ability to make design decisions within architectural projects while demonstrating synthesis of user requirements, regulatory requirements, site conditions and accessible design and consideration of *measurable environmental impacts* of their design decisions. (NAAB 2020)

This last criterion must be demonstrated in the passing design work of *all* students preparing to enter practice. My introduction of the phrase *measurable environmental impacts* was resisted by some at the time since it did not mention carbon footprinting by name and also because it was unclear how design students would measure the broad environmental impacts of design alternatives since they had never had to before. It was finally agreed that quantification of environmental impacts would be included as a student requirement. If the two preceding accreditation conditions that centered on environmental stewardship, ecological knowledge, and professional responsibility were to *actually* support public health, safety, and welfare, then each and every student about to enter the profession had to prove how his or her design decisions consider and act on the insights the *non-geometric* environmental impact category data can reveal.

James Fava ended his chapter in 2012 saying that "the jury was still out on whether this trend will continue" but that he predicted that LCA information would be used more and more to inform decision-making in both government and private sectors. It seems that he, and all who have been optimistic about LCA's ability to augment many other fields, was right. They predicted wider operationalizing of LCA data through new visualization tools (Fig. 6.3).

The last three chapters of this book focus on a student-centric approach to build broader awareness and encourage the use of several LCA/LCIA data interfaces. Several approaches were evaluated, from semiautomatic processes that leverage building or product information models such as Tally, One Click LCA, and Solidworks Sustainability to spreadsheet processes that work off of the Athena Impact Estimator for Buildings. Much of this work came out of a close collaboration with an independent study undergradaute architecture student, Erin Heidelberber. The work is primarily focused on the building scale due to the majority of design tools that can interface with building centric data. Putting these practical approaches into practice, designers of every stripe can augment their creative impulses and improve judgment with a growing number of effective LCIA tools that eliminate much of the painstaking work required in a proper quantitative life cycle approach to design.

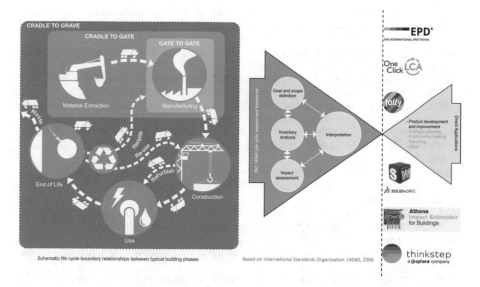

Fig. 6.3 A growing number of LCA/LCIA tools are available to designers to reduce friction in sustainable design decision-making

References

Arrhenius S (1896) On the influence of carbonic acid in the air upon the temperature of the ground. Philos Magaz J Sci 41:237–276

Björnberg KE, Karlsson M, Gilek M, Hansson SO (2017) Climate and environmental science denial: a review of the scientific literature published in 1990–2015. J Clean Prod 167:229–241. https://doi.org/10.1016/j.jclepro.2017.08.066

Cays J (2017) Life-cycle assessment: reducing environmental impact risk with workflow data you can trust. Archit Des 87:96–103. https://doi.org/10.1002/ad.2179

Cook J, Nuccitelli D, Green SA et al (2013) Quantifying the consensus on anthropogenic global warming in the scientific literature. Environ Res Lett 8:024024. https://doi.org/10.1088/1748-9326/8/2/024024

Cook J, Oreskes N, Doran PT et al (2016) Consensus on consensus: a synthesis of consensus estimates on human-caused global warming. Environ Res Lett 11:048002. https://doi.org/10.1088/1748-9326/11/4/048002

Drabek T (1986) Human system responses to disaster: an inventory of sociological findings. Springer Series Environ Manag 72. https://doi.org/10.1007/978-1-4612-4960-3

Elkington J (1998) Cannibals with forks: the triple bottom line of 21st century business. New Society Publishers, Gabriola Island, BC/Stony Creek, CT

EUR 24708 EN - Joint Research Centre - Institute for Environment and Sustainability (2010) ILCD handbook: general guide for life cycle assessment: detailed guidance. Publications Office of the European Union, Luxembourg

Fava J (2012) Life cycle knowledge informs greener products. In: Life cycle assessment handbook: a guide to environmentally sustainable products, 1st edn. Wiley-Scrivener, Hoboken, pp 585–596

Frischknecht R, Jungbluth N, Althaus H-J et al (2005) The ecoinvent database: overview and methodological framework (7 pp). Int J Life Cycle Assess 10:3–9. https://doi.org/10.1065/lca2004.10.181.1

Grabe ME, Zhou S, Barnett B (2001) Explicating sensationalism in television news: content and the bells and whistles of form. J Broadcast Electron Media 45:635–655. https://doi.org/10.1207/s15506878jobem4504_6

Hannson SO (2017) Science denial as a form of pseudoscience. Stud Hist Phil Sci 39:39–47. https://doi.org/10.1016/j.shpsa.2017.05.002

Kahneman D (2013) Thinking, fast and slow, 1st edn. Farrar, Straus and Giroux, New York

Laibson D (1997) Golden eggs and hyperbolic discounting. Q J Econ 112:443–478

Lambert C (2006) The Marketplace of Perceptions. In: Harvard Magazine. https://harvardmagazine.com/2006/03/the-marketplace-of-perce.html. Accessed 29 Sep 2020

Murakami H, Vecchi GA, Delworth TL et al (2017) Dominant role of subtropical Pacific warming in extreme eastern Pacific hurricane seasons. J Clim 30:243–264. https://doi.org/10.2307/26387480

NAAB (2020) Conditions for accreditation. National Architectural Accrediting Board, Inc, Washington, DC. https://www.naab.org/wp-content/uploads/2020-NAAB-Conditions-for-Accreditation.pdf

NASA (2020) Do scientists agree on climate change? In: Climate Change: change: vital signs of the planet. https://climate.nasa.gov/faq/17/do-scientists-agree-on-climatechange

Nawaguna Their Districts Are at Risk. But They Still Vote 'No' on Climate Action – Roll Call. https://www.rollcall.com/2018/10/17/their-districts-are-at-risk-but-they-still-vote-no-on-climate-action/. Accessed 4 Aug 2020

Nickerson RS (1998) Confirmation Bias: a ubiquitous phenomenon in many guises. Rev Gen Psychol 2:175–220

NW 1615 L. St, Suite 800Washington, Inquiries D 20036USA202–419-4300 | M-857-8562 | F-419-4372 | M (2020) Two-Thirds of Americans Think Government Should Do More on Climate. In: Pew Research Center Science & Society. https://www.pewresearch.org/science/2020/06/23/two-thirds-of-americans-think-government-should-do-more-on-climate/. Accessed 24 Jul 2020

Oreskes N, Conway EM (2011) Merchants of doubt: how a handful of scientists obscured the truth on issues from tobacco smoke to global warming, Reprint edition. Bloomsbury Press, New York

Rebitzer G, Ekvall T, Frischknecht R et al (2004) Life cycle assessment: Part 1: framework, goal and scope definition, inventory analysis, and applications. Environ Int 30:701–720. https://doi.org/10.1016/j.envint.2003.11.005

Schwartz N, Bless H, Strack F et al (1991) Ease of retrieval as information: another look at the availability heuristic. J Pers Soc Psychol 61:195–202

Simonen K (2014) Life cycle assessment, 1st edn. Routledge, London /New York

Smil V (2008) Global catastrophes and trends: the next fifty years. The MIT Press, Cambridge, MA

Tyson A, Kennedy B (2020) Two-thirds of Americans think government should do more on climate. Pew Research Center Science & Society. Washington, DC. https://www.pewresearch.org/science/2020/06/23/two-thirds-of-americans-think-government-should-do-more-on-climate/

US EPA O (2016) Understanding global warming potentials. In: US EPA. https://www.epa.gov/ghgemissions/understanding-global-warming-potentials

Van Hoeck N, Watson PD, Barbey AK (2015) Cognitive neuroscience of human counterfactual reasoning. Front Hum Neurosci 9:420. https://doi.org/10.3389/fnhum.2015.00420

Weidema BP (1998) Multi-user test of the data quality matrix for product life cycle inventory data. Int J Life Cycle Assess 3:259–265. https://doi.org/10.1007/BF02979832

Chapter 7
LCI Data and Design

Abstract LCA gathers and presents facts that can appeal to both those who aspire to save the planet while not materially impacting the financial bottom line.

Steady and relentless fact-based reporting can counteract the worst effects of climate denialism. Data-driven actions are especially effective when aligned with major stakeholder interests. LCA systems thinking is predicated on the quality and quantity of available foundational *life cycle inventory* (LCI) information. It is transparent in how it arrives at its conclusions and qualifies its findings. Thus, even a result with a relatively low level of certainty, when accurately stated, is superior to an unquantified assertion.

LCA data quality improvements in recent years create the foundation for a trustworthy evaluation system. More public and private service providers are leveraging statistically more accurate environmental impact reporting. Verified quality data provides a continuity that compensates for the vagaries of changing political decisions such as the funding or defunding of any particular nation's environmental protection agency.

When armed with "good" data, designers are in a unique position to present options, early in decision-making stages, that simultaneously provide elegant solutions and improve the environmental profile of a project. Multiplying these effects by all of the projects of even a single designer who employs LCA over the course of a career creates a material environmental benefit. This chapter presents an overview of the LCI data and the environmental impact assessment tools available to designers working at the building scale.

7.1 Designers Engaging with LCI Data

Following the initial study boundaries laid out in the goal and scope phase, the LCA expert turns her attention to life cycle inventory (LCI) data. Ensuring the quality of LCI data is critical to the validity and success of any LCA study, as data quality and relevance directly impact the results of a study (Saade et al. 2018). However, it is also not something that the average designer will typically engage with. It is one of the hardest and longest parts of conducting a full LCA on even a small single product system and requires a much different way of thinking and skill set than designers are typically trained in Curran (2012). Furthermore, the sheer volume of data

© Springer Nature Switzerland AG 2021
J. Cays, *An Environmental Life Cycle Approach to Design*,
https://doi.org/10.1007/978-3-030-63802-3_7

required for a study on a building assembly, or whole building, which combines a vast array of materials over their life spans, from raw material extraction to end of life, is a prohibitive task (Van Ooteghem and Xu 2012).

Formal LCA phase 2 life cycle inventory activities include three main procedures. The first is data collection, the second is data calculation, and the third is the determination of how the various system inputs and outputs are to be allocated, if necessary, between different products and functions. The LCA operator will validate that all unit processes obey fundamental laws of mass and energy conservation throughout the process. Detailed graphic flow charts show how each unit process relates to the functional unit established in phase 1. (See Chap. 5 for function unit description.) Physical input and output data must be appropriately aggregated based on relationships to the study goal, equivalent individual substances, and equivalent environmental impacts.

The first (data collection) and second (data calculation) procedures produce a clearer, second iteration, data picture that is used to refine the original product system boundary in phase 1. Sensitivity analyses performed on the calculated data are used to exclude insignificant life cycle stages or unit processes or to include new ones shown to be significant. The third procedure provides either the basis to avoid or the justification for allocating physical inputs and outputs within a bounded product system (ISO 14044 2006b).

Prior to carrying out LCI data collection, calculation, and allocation activities, the LCA expert must identify all primary physical material extractions required for and emissions released in every product life cycle stage and decide if they will use an economic input/output or detailed process-based inventory analysis to calculate (Jolliet et al. 2015).

Fortunately for the non-specialist, there exist a number of different certified LCI databases that have their own processes and standards to verify data quality. These databases allow the designers to avoid much of the unfamiliar and time-consuming process of validating every data source while still seeing the overall criteria that support the LCI data. These databases feed into impact assessment (LCIA) tools design workflows to improve decision-making. Several programs designed to bring LCA into the hands of designers match existing LCI data values to material takeoffs drawn from a digital geometric design model. This allows designers to relatively quickly assess the environmental profile of their design without needing the knowledge and time to carry out a full LCA study. This removes the designer from the details concerning LCI data collection, calculation, and allocation processes. Instead, these programs draw data from existing open public or proprietary datasets.

7.2   Use of LCI Databases in LCIA Tools

One way that designers engage with LCI data, without even realizing it, is through the use of LCIA tools. Several LCI databases directly fuel the various LCIA tools available to designers, explored in depth in Chap. 8. However, this all happens as background processes, and the designer is completely removed from the process. Instead, users enter information about material quantities and receive information in the LCIA phase of the design. The whole LCI phase happens completely outside of the control and view of the user. This does not make this stage in the process any less important. In using an LCI database, and by removing the user from the LCI phase of the study, LCIA tools make the process much faster and accessible to a wider audience.

While the Athena Impact Estimator for Buildings and the Building Industry Reporting and Design for Sustainability (BIRDS) framework both operate off proprietary databases, there is overlap in the data sources for the rest of the programs discussed in this book (Fig. 7.1). Several notable LCI databases, including GaBi, ecoinvent, and USLCI, feed the LCIA and LCA tools discussed in Chap. 8. All of the databases operate using process-based data as opposed to input-output (IO), except for the National Institute of Standards and Technology (NIST) database which combines the two. In a process-based LCI database, the inventory data is calculated based on measured flows of matter, energy, and pollutants, as opposed to the IO approach which equates money spent in broad economic sectors to environmental impacts (Jolliet et al. 2015). Process-based data can be obtained from government databases, published industry averages, journals or reference books, and specific manufacturing practices and by physically measuring flows of matter and electricity at manufacturing sites (Curran 2012). The LCI data of a designated unit process consist of all of these reference flows.

Figure 7.1 makes a distinction between LCIA and LCA tools, which are connected to the design workflow in the next chapter first in Sect. 8.1. However, there

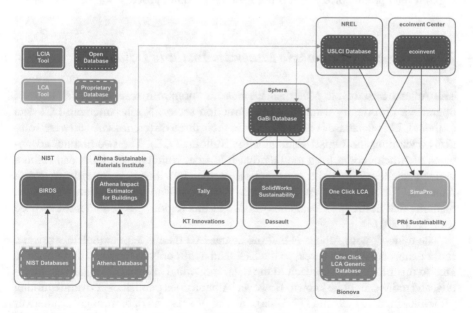

Fig. 7.1 Partial data sources of prominent LCIA/LCA software for designers in North America

is also a distinction made between proprietary and open LCI databases. The GaBi, ecoinvent, and USLCI databases are all open. The USLCI database is fully open and free to use, while GaBi and ecoinvent require a license fee to access. However, all of the information about the data are presented to the user. All of the individual reference flows that make up the unit process, in addition to the metadata surrounding when and how the data were collected, are freely accessible to the user. These databases can be used on their own or through one of the many LCIA/LCA tools available. The proprietary databases, on the other hand, are kept fully private. Users can never gain access to the LCI data or metadata contained in the database. Rather, this information can only be used through an LCIA tool where the user is completely removed from the LCI process.

The NIST databases, Athena databases, and USLCI databases are all specific to North American conditions, but the GaBi, ecoinvent, and One Click LCA databases all contain global data. Geographic correlation of data to the location of the study can impact the results based on material availability, manufacturing practices, grid mixes for energy used in manufacturing or during the use phase, transportation distances, and end-of-life conditions. The most reliable data for a study are from the area of study, or representative of the averages across a large region that encompasses the area of study; however, data from an area with similar manufacturing processes can also be relevant and applicable to a study (Ciroth et al. 2016). Using data from another region is not always wrong, but it does introduce more uncertainty to the study. Data from a different region is often used as proxy data if the information about a specific reference flow in the region of the study is missing (Hester et al. 2018). Based on the very time- and cost-intensive process of generating new data points, proxy data is used for background processes in many studies.

7.2.1   *Athena Sustainable Materials Institute Database*

The Athena Sustainable Materials Institute, a "nonprofit research collaborative," operates a proprietary database of construction sector, North American LCI data (Athena). This database is used directly in their three different LCA software solutions, including the Impact Estimator for Buildings (IE). The two distinct advantages of Athena come from the fact that it is free, which means LCA can reach a wider audience within the construction sector, and that it is specific to North America. The vast majority of LCI data is concentrated on European conditions, meaning that in many LCA studies outside of this region, proxy data must be used (Suh et al. 2016).

The tradeoff with Athena is that the actual LCI data is kept completely private. Individuals cannot gain access to the LCI data at all, only to the tools that utilize it. Due to this black-box approach to the data, individual criteria, dates, testing methods, and more cannot be known. However, Athena does publish their overall guiding principles on data collection (Athena). Instead of sourcing data from government or trade sources, all of the information is built using actual manufacturing models. It is representative of industry averages for each specific material and is regionally sensitive in regard to energy grid mixes, recycling practices, manufacturing tech-nologies, and transportation distances. Furthermore, beyond individual products, the database includes information on "energy use, transportation, construction and demolition processes including on-site construction of a building's assemblies, maintenance, repair and replacement effects through the operating life, and demoli-tion and disposal" (Athena n.d.). The reputation of the Athena Institute helps to instill trust in the data even with the complete lack of transparency.

7.2.2   ecoinvent Database

Taking an opposite stance to that of Athena, the ecoinvent database was built on the principles of developing a transparent and consistent set of LCI data spanning many economic sectors (Frischknecht et al. 2005). Originally this data was specific to Switzerland, where it is developed and managed, but with the release of version 3, significant effort has been made to expand the database to encompass global LCI data (Saade et al. 2018). Anyone with a license to the database can see all of the values of the individual flows within the unit process, all of the metadata associated with the dataset, the calculated level of uncertainty for all flows within the unit pro-cess, and an assessment on the data quality and relevance using an LCI pedigree matrix (Frischknecht et al. 2005). ecoinvent has a very systematic method for calcu-lating uncertainty of the data that was updated in 2016 with the release of version 3.0 (Ciroth et al. 2016). The data is updated yearly to add new datasets, update the existing ones, and make general improvement to the database. As of version 3.6, there are around 17,000 LCI datasets included in the database (ecoinvent). Additionally, ecoinvent provides the user with options regarding impact partition modeling and LCIA methods (Saade et al. 2018). The meticulous, transparent, and customizable nature of ecoinvent makes it one of the most widely used LCI data-bases in the world.

The transparency outlined above is valuable for LCA experts involved in the LCI phase of a study. However, designers will typically only use the data through an LCIA tool and therefore will not have the opportunity to engage as closely with all of these features, nor do they have the requisite background knowledge and skills to engage with them in any meaningful way. Instead, designers only have as much information about the data as is available to them through the specific tool that is being used. For more information on this, see Sect. 7.3.

7.2.3   GaBi Database

Much like the ecoinvent database, the GaBi database is also focused on providing consistent and transparent LCI data across sectors and geographical regions. Sphera, formerly thinkstep, owns and operates the database. The data they include in their database is built off real industrial practices, not theory or concepts. In addition to owning and managing their own datasets, GaBi also licenses data from ecoinvent and the USLCI database (Sphera). The flow hierarchies of these databases are matched to the one used in GaBi to ensure seamless integration. The database itself houses tens of thousands of datasets spanning the majority of industries including agriculture, building and construction, metals and mining, plastics, textiles, and many more.

The quality of the data is ensured through frequent updates and external reviews. Every dataset in the database is updated annually to best match current and changing practices. Furthermore, Sphera contracts Dekra to conduct a third-party critical review of the LCI data to ensure both quality and accuracy (Sphera). Beyond the attention paid to quality, transparency is a critical component of the database. All of the metadata associated with the datasets are available freely online, even without having a license to access the database. For each individual process, a host of information is available, that provides users with the ability to assess the relevance of the data to a particular study. The information provided includes geographical location, reference year and year it is valid until, general information, reference flows and a flow diagram, technology description, methodologies, allocations, completeness, and the data quality indicator based on Ciroth's pedigree matrix. Since this is freely posted on their website, even designers who are using the LCI data through a tool and not licensing the database can investigate the data and determine the quality and reliability as it relates to the goal and scope of the study.

7.2.4 USLCI Database

Started in 2003 with the aim to provide high-quality, transparent LCI data relevant to US conditions to support the expanded use of LCA studies in decision-making, the USLCI database is recognized as the main source of LCI data in the United States (NREL 2009). The database is maintained and managed by the National Renewable Energy Laboratory (NREL). The information in the USLCI database is free to access and use (Curran 2012). In the database itself, both the data and metadata are transparently reported. The datasets include information on the reflected manufacturing process, the actual flows of materials and emissions, and any applicable allocation factors. There is also information on who generated, documented, and published the data, when the data was published, and how the data was obtained.

Efforts have been made to not only keep the data up-to-date but also to provide a wide range of unit processes and flows in different sectors to provide users with a comprehensive set of data that enables a consistent analysis (NREL 2009). While some datasets are listed geographically as "North America," all of the datasets are most representative of US conditions having been built on US data and modeling technology (Grabowski et al. 2015). However, major efforts have been made to keep the database compatible with international LCI databases, and NREL has sought help from many international players to do so (NREL 2009). This is important as multiple datasets are often needed to complete a single study (Suh et al. 2016). Overall, the USLCI database freely provides transparent information to users based on US conditions.

7.2.5 NIST Databases

BIRDS is fed by three proprietary databases owned and operated by NIST. There are separate databases for single-family residential and commercial buildings as well as one that explores incremental energy efficiency measures based on data gathered from NIST's Net-Zero Energy Residential Test Facility (NZERTF) (Kneifel and O'Rear 2018). BIRDS explores LCA, LCC, operational energy, and indoor environmental quality, and therefore these databases include more information than just LCI data. Specific to the LCI data, most of the data on building components are based on a mix of top-down and bottom-up data as it is available. More specific process-based data is used for the impacts of operational electricity and natural gas as well as any of the building's energy technologies as BIRDS focuses largely on the use phase of the building (Kneifel et al. 2018b). In 2018, the data were updated and expanded to help ensure the validity of any studies carried out using the data (Kneifel and O'Rear 2018). Much like the Athena database, everything is kept completely private, and no transparency on the data is offered.

7.3 Transparency of LCI Data to Designers

In a 2003 survey conducted on users of the Building for Environmental and Economic Sustainability (BEES) software, the majority of responses indicated that transparency in all stages of using the tool is important (Hofstetter, Mettier 2003). However, this transparency can come at the burden of additional time and decisions from the designer. Nevertheless, the inclusion of transparency surrounding LCI data is critical in designers trusting the tools available to them. Without updated LCI data, the tools operating on them become obsolete (Kneifel and O'Rear 2018). The tools using LCI data are only as reliable as the data itself is. Each of the tools, explored in Chap. 8, has a different way of handling LCI data transparency.

7.3.1 Athena Impact Estimator for Buildings

As discussed in Sect. 7.2.1, Athena has very limited transparency surrounding their LCI data. Limited insight into the actual flows of material can be gathered from the results of a study completed using the IE software. While the results are by and large presented for the impact assessment, the excel report starts to break down individual flows of materials. Energy consumption, air emissions, water emissions, land emissions, and resource use are all broken down into individual components, and their total flows are calculated and tabulated. These results, which come directly from the LCI data, are presented by discrete life cycle stages, but the sum of flows for all materials is presented together, not separated by individual material. However, this can only be viewed after all of the material quantities and assemblies have been entered and is not readily available from within the interface. Furthermore, it does not give any information on when or how the data was collected. Overall, Athena provides extremely limited insight into the LCI data.

7.3.2   BEES

The LCI data for the Building for Environmental and Economic Sustainability (BEES) tool, an LCIA and LCC tool at the level of building products, is derived from both generic averages and manufacturer-specific products (Kneifel and O'Rear 2018). This comes from US average data, unit process and facility-specific information, and published EPDs (Kneifel et al. 2018a). Transparent information on the specific datasets used in BEES can be found online but not in the tool. The products section of the online documentation of BEES provides a list of all of the products that BEES currently has information on. For each of these products, there exists a document containing information on the manufacturer, if the data is about a specific product, as well as the system boundaries, raw materials, manufacturing, transportation, use, end of life, and sources for the data. This is all accessible for the user to evaluate the applicability of specific datasets to a study, just not integrated into the tool itself.

7.3.3   BIRDS

While the source of the LCI data used in the BEES framework is reported online, the LCI data used in BIRDS lacks transparency. Little information about the databases themselves, discussed in Sect. 7.2.5, is published, and the user has no information on what the LCI data encompass nor any of the metadata surrounding it.

7.3.4   Tally

Tally LCI data is sourced from the GaBi database, discussed in Sect. 7.2.3. In the Tally dashboard itself, minimal information about the data is displayed, but more information can be found in the generated report. A box on the side of the interface provides quick insight into the manufacturing processes and material quantities that the data represents. Furthermore, any data that is based on a specific manufacturer is explicitly labeled as so, but no metadata for any of the specific entries is provided. In the PDF report generated by Tally, a note about data quality is included. "Data quality is judged by its measured, calculated, or estimated precision; its completeness, such as unreported emissions; its consistency, or degree of uniformity of the methodology applied on a study serving as a data source; and geographical, temporal, and technological representativeness." Later on in the same PDF, the individual LCI sources are listed by name with the region and date of the data source. If it is based on an EPD, the number, program operator, and expiration date are also listed.

7.3.5   *SolidWorks Sustainability*

SolidWorks Sustainability also sources LCI data from the GaBi databases, discussed in Sect. 7.2.3. There is no transparency regarding the data built into the interface or any of the reporting measures. SolidWorks Sustainability requires more information on the specific materials, manufacturing processes, and locations, so therefore the LCI data that the program pulls will be relevant to the part or assembly being analyzed. However, none of the specifics of the LCI datasets or accompanying metadata are displayed to the user.

7.3.6   *One Click LCA*

One Click LCA is the only LCIA tool explored in this book that builds transparency about LCI data directly into the tool. Next to every material definition in both the Revit and web interface is a button that reveals all of the metadata about the LCI data source (Fig. 7.2). This includes the database it comes from, the country it was completed in, the year it was completed, what standards it follows, what EPD and PCR it is based on (if any), and other relevant information about the LCI data. This level of specificity helps to instill additional trust in the user as well as improve the appropriateness of the data being used and therefore increase the accuracy of the overall study. Specifics regarding data collection, boundary conditions, allocation factors, and measured reference flows are not available to the user. Only general metadata is visible. One Click LCA sources LCI data from all over, and the data sources vary based on the location of the study and what license the user has (Bionova n.d.). Among other databases, the information comes from GaBi, ecoinvent, and USLCI. One Click LCA also maintains their own generic construction material database. More information on LCI data used in One Click LCA can be accessed   at   https://www.oneclicklca.com/support/faq-and-guidance/documentation/database/.

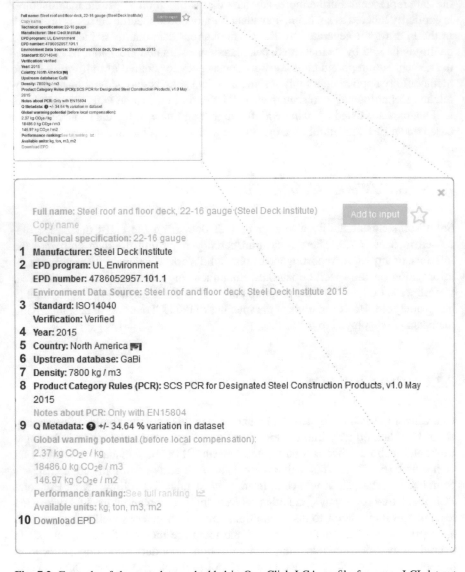

Fig. 7.2 Example of the metadata embedded in One Click LCA profile for every LCI dataset assigned to project materials. These provide key source information on:

Manufacturer—specific manufacturer or organization that produced the data point

Year—when the study was conducted that produced the data

Country—what applicable country or region for data point

Upstream database—specific database source of the data point

EPD program and number—an EPD on which a data point is based if any. Provides a direct link for users to further evaluate the EPD source PCR—product category rules that govern the data point

Q Metadata—variability in the data point based on manufacturer, product, and plant specificity used to gather LCI data. This is based on the Carbon Leadership Forum methodology

Verification and Standard—confirmation that the data point has been verified and lists any relevant standards to which the data point conforms

7.3.7 SimaPro and Other LCA Tools

In this chapter SimaPro is being used as an example of an LCA tool, as opposed to the LCIA tools discussed above. SimaPro also sources data from many different places, including the ecoinvent and USLCI databases, and the sources encompass both process-based and IO data (PRé). More information on LCI data in SimaPro can be found at https://simapro.com/databases/. The GaBi LCA software uses the GaBi database, which gets information from the USLCI and ecoinvent databases (Sphera). There is significant overlap in the data that services LCA tools aimed at experts and LCIA tools aimed at designers, showing that these studies are built on the same reliable LCI foundations.

7.4 Agreement in LCI Data

Across the databases, there is a lot of overlap between the data sources and specific LCI data (Reis 2013). One example of this is the way that the GaBi database and the USLCI database both feed into each other, which causes significant overlap in data. Beyond that, many primary sources of the data are the same leading to very similar, if not identical, data points. New databases are evaluated based on their general agreements with existing, verified databases (Jolliet et al. 2015). From an outside perspective, skepticism in the LCI data and overall LCA process can be a barrier to designers adopting this information flow into their practices (Saunders et al. 2013). However, an understanding of the rigor taken in collecting and ensuring the databases, in addition to the agreement across them, is a sound foundation on which designers can build trustworthy workflow.

References

Athena Sustainable Materials Institute (n.d.) www.athenasmi.org. Accessed June 2020
Bionova (n.d.) One Click LCA. www.oneclicklca.com. Accessed June 2020
Ciroth A, Muller S, Weidema B et al (2016) Empirically based uncertainty factors for the pedigree matrix in ecoinvent. Int J Life Cycle Assess 21:1338–1348. https://doi.org/10.1007/s11367-013-0670-5
Curran MA (ed) (2012) Life cycle assessment handbook: a guide for environmentally sustainable products. Scrivener Publishing\Wiley, Salem\Hoboken, pp 105–106, 115
ecoinvent Centre (n.d.) https://www.ecoinvent.org. Accessed July 2020
Frischknecht R, Jungbluth N, Althaus HJ, et al (2005) The ecoinvent database: overview and methodological framework. Int J Life Cycle Assess 10:3–9. https://doi.org/10.1065/lca2004.10.181.1
Grabowski A, Selke SEM, Auras R et al (2015) Life cycle inventory data quality issues for bioplastics feedstocks. Int J Life Cycle Assess 20:584–596. https://doi.org/10.1007/s11367-015-0853-3
Hester I, Miller TR, Gregory J et al (2018) Actionable insights with less data: guiding early building design decisions with streamlined probabilistic life cycle assessment. Int J Life Cycle Assess 23:1903–1915. https://doi.org/10.1007/s11367-017-1431-7

Hofstetter P, Mettier TM (2003) What users want and may need: insights from a survey of a life-cycle tool. J Ind Ecol 7:79–101. https://doi.org/10.1162/108819803322564361

ISO (2006b) ISO 14044 Environmental Management: Life Cycle Assessment: Requirements and Guidelines. International Organization for Standardization, Geneva

Jolliet O, Saade-Sbeih M, Shaked S et al (2015) Environmental life cycle assessment. CRC Press, Boca Raton, pp 47–66

Kneifel J, O'Rear E (2018) Challenges and opportunities in quantifying and evaluating building sustainability. Technol: Archit + Des 2:160–169. https://doi.org/10.1080/24751448.2018.1497363

Kneifel J, Greig AL, Lavappa P, et al (2018a) Building for Environmental and Economic Sustainability (BEES) Online 2.0 technical manual. National Institute of Standards and Technology. https://nvlpubs.nist.gov/nistpubs/TechnicalNotes/NIST.TN.2032.pdf. Accessed July 2020

Kneifel J, O'Rear E, Webb D, et al (2018b) Building Industry Reporting and Design for Sustainability (BIRDS) Building code-based residential database technical manual: update. National Institute of Standards and Technology. https://nvlpubs.nist.gov/nistpubs/TechnicalNotes/NIST.TN.1999.pdf. Accessed July 2020

National Renewable Energy Laboratory - NREL (2009) U.S. LIFE CYCLE INVENTORY DATABASE ROADMAP. US Department of Energy. https://www.nrel.gov/docs/fy09osti/45153.pdf. Accessed July 2020

PRé Sustainability (n.d.) SimaPro. https://simapro.com/. Accessed July 2020

Reis L (2013) An exploration of materials taxonomies to support streamlined life cycle assessment. Dissertation, Massachusetts Institute of Technology

Saade MRM, Gomes V, Silva MG et al (2018) Investigating transparency regarding ecoinvent users' system model choices. Int J Life Cycle Assess 24:1–5. https://doi.org/10.1007/s11367-018-1509-x

Saunders CL, Landis AE, Mecca LP et al (2013) Analyzing the practice of life cycle assessment: focus on the building sector. J Ind Ecol 17:777–788. https://doi.org/10.1111/jiec.12028

Sphera (n.d.) GaBi Solutions. www.gabi-software.com. Accessed July 2020

Suh S, Leighton M, Tomar S et al (2016) Interoperability between ecoinvent ver. 3 and US LCI database: a case study. Int J Life Cycle Assess 21:1290–1298. https://doi.org/10.1007/s11367-013-0592-2

van Ooteghem K, Xu L (2012) The life-cycle assessment of a single-storey retail building in Canada. Build Environ 49:212–226. https://doi.org/10.1016/j.buildenv.2011.09.028

Chapter 8
Dashboard

Abstract Evolving LCA interfaces simplify and translate complex data for non-specialists. Intuitive, visually compelling signals inform judgment and improve outcomes. Seeing clearly guides action.

The dashboard has evolved in its primary purpose as a protective device. Initially, dashboards were purely physical objects, planks of wood placed between the carriage and galloping horses to block the "dashing up" of mud, rocks, water, and other debris onto travelers. That early functionality has developed into a locus of increasingly complex, critical information about the current state of dynamic machines, systems, and networks. Evolving hardware and software interfaces increase the access, understanding, and effectiveness of environmental impact data. This, metaphorically, returns the dashboard to its original, protective role for those capable of "harnessing" the flood of information in real time.

Sustainable dashboard information focusing on life cycle environmental impact assessment guides improved decision-making. Designers familiar with and capable of navigating this dashboard information are better positioned to make a case for their own design solutions.

Designers can use tools to satisfy final customer demands for sustainable products and services. Individual consumers are not the target market for existing LCA tools. BEES, ATHENA, EIE, Tally, and SolidWorks Sustainability put LCA in the hands of designers, who translate the information and present cohesive, data-driven options to clients.

8.1   The LCIA Dashboard

A full LCA process can answer the very important questions of the impacts a design has on the environment. A designer reading through study conclusions or published EPDs can potentially utilize this information to make informed judgments in a design project or any material subcomponent of it. These documents form a framework to provide designers with qualified facts and information they can use in their decision making process. However, no LCA document can, by itself, provide designers with a single explicit answer or course of action. LCI and LCIA data processed through the lens of an integrative design software interface is currently the

© Springer Nature Switzerland AG 2021
J. Cays, *An Environmental Life Cycle Approach to Design*,
https://doi.org/10.1007/978-3-030-63802-3_8

most effective way to augment a design workflow to yield supportable insights on environmental impacts throughout a system's life cycle.

8.2   Life Cycle Impact Assessment and Interpretation

After collecting all of the LCI data, the next step for an LCA practitioner is to move into the impact assessment phase of the study as defined in ISO 14044, if this is part of the goal and scope of the study. Non-geometric impact assessment results from numerous studies on each material are embedded in software packages augmenting geometric design modelers and are invisible to the designer. A schematic understanding of these quantitative techniques, however, can provide insight to the non-specialist and increase the confidence in the use of software that can improve judgment regarding sustainable design decisions.

There are numerous absolute evaluative techniques such as environmental performance evaluation, environmental impact assessment, and risk assessment that may be used to inform the relative LCIA results that revolve around the user-defined functional unit and the initial goal and scope of the study. Anchoring results to the functional unit maintains the internal logic of the system study, helping to evaluate and manage uncertainty caused by data quality issues, boundary cutoff decisions, and the relevance of the results that can decrease depending on the way the LCI functional unit is calculated, how results are averaged, aggregated, and allocated within the entire product system.

Impact assessments are provided for each impact category (see Chap. 5) that is material to the study. The LCA expert selects from among many midpoint categories such as climate change, acidification, eutrophication, abiotic depletion, stratospheric ozone destruction, and endpoint damage categories such as human health, natural resources/ecosystem services, ecosystem quality, and anthropogenic impacts and assigns environmentally relevant quantifiable category indicators, for example, the amount of infrared radiative forcing expressed in W/m^3 caused by the greenhouse emissions created within a product system.

Characterization models such as the IPCC baseline model of 100-year timeframe are used to produce characterization factors (CFs) used to convert the raw LCI data into common units that can be aggregated and applied within an impact category. LCIA research continues to evolve, data continue to multiply, and accuracy in impact category indicators continue to improve (Hauschild et al. 2013).

Several non-mandatory elements can also be used to inform and qualify the impact assessment. Normalization of category indicator results against a baseline reference indicator is one. Impact category grouping after sorting and subjectively ranking them based on priorities important to an individual or group is another. Subjectively weighting results is a third. In every case, indicator results prior to any of these optional processes is ideally provided to ensure transparency and adherence to generally accepted reporting protocols.

Differences in qualitative value choices, variation in precision among impact categories among other LCIA methodological limitations, preclude it from acting as the sole basis of publicly disclosed comparative assertions of environmental superiority of one specific product over another. But LCIA is useful in general large-scale industrial sector product descriptions that can be evaluated against one another especially if this is the stated purpose of the study.

In the LCIA phase, all of the individual inventory flows are classified into the several environmental impact categories that they contribute to. For example, several gasses beyond carbon dioxide, including methane and nitrous oxide, contribute to GWP (Kneifel et al. 2018). After each flow is assigned to an impact category, a characterization factor (CF) must be applied to all of the flows in one category to get a single equivalency impact score. These CFs reduce down a list of thousands of flows into a handful of impact categories so that the results of the study can be more easily understood.

The results of the LCIA phase can be presented as midpoint or endpoint indicators. Midpoint indicators combine the flows into equivalent values of one reference flow (De Schryver 2011). Again using the example of GWP, kg CO_2 is commonly used to represent the GWP of all chemicals, not just carbon dioxide. Endpoint indicators attempt to trace the potential impacts to the end of the cause-effect pathway and quantify the damage done to human health, ecosystem health, and resource availability (De Schryver 2011). While endpoint indicators can be more illustrative or easier to comprehend on a quick glance, they also introduce more subjectivity into the study, whereas the uncertainty of midpoint indicators is much lower (Curran 2012). All of the tools presented in this chapter primarily utilize midpoint indicators.

8.2.1 A Note on Interpretation

Interpretation is the fourth but iteratively repeating phase that specialists must complete when conducting an LCA study. It requires professional judgment to interpret multiple results produced over successive passes through the other three phases. All results from the LCI and LCIA phases are evaluated against the framing system functions, functional unit, and system boundaries defined in the goal and scope phase. The LCA operator will identify and report limitations and increased uncertainty created by data quality issues based on completeness, consistency, and sensitivity checks. If there is any missing or incomplete information, another pass through the other phases is required. Once all constituent parts are reasonably complete and determined to be consistent with and supportive of the goal and scope of the study, the LCA professional presents final qualified conclusions and recommendations to the individual or group who commissioned the study.

8.2.2 LCIA Tools for Designers

These dynamic models and the databases of interpreted environmental impact information that flows from them feed design software dashboards. LCIA tools aimed at designers and other non-LCA experts take steps to simplify the otherwise rigorous, knowledge-intensive, and time-consuming process of completing an LCA study. In order to do so, many steps of the process happen behind the scenes without any input from the users. In most cases, the tools are primarily focused on the LCIA phase, with limited input on other stages of the process. Users are required to enter a small amount of information pertaining to the goal and scope, and the role of LCI data in these tools is discussed at length in Chap. 7. Otherwise, users enter basic information about material selection and quantities and get the LCIA results back from the tool. The decision to distance these tools from the early stages of the assessment process is both disadvantageous and beneficial. This decreases understanding and transparency in LCA studies and increases discrepancies in assessment methods making it much harder to compare results across studies (Rodriguez et al. 2019). Only studies where all methodological details are consistent can be compared. However, this also makes the tools easier and less time-consuming to use meaning that more designers can adopt them into their workflows without being an LCA expert.

LCIA tools serve as a window into LCA studies for designers and other non-experts to improve their decision-making. An LCIA tool differs from an LCA tool by distancing the user from most of the steps in a formal LCA study. In an LCIA tool, the user adds information about geometry or material quantities and material definitions and gets the LCIA results. An LCA tool brings the user into methodological decisions and the other stages of the process. Designers lack the requisite knowledge to carry out a full LCA study, but LCIA tools bring the complex, scientific process of LCA to a wider audience including designers. Without them, it would be much more resource, knowledge, and time extensive to bring measured change to a project.

8.3 Impact Assessment Methods

All of the outputs that are combined within one impact category have similar impacts on the environment and therefore are like to like comparisons. However, there are still differences and subjectivity involved in how the data is transferred from the LCI phase to the LCIA phase.

Several widely accepted impact assessment methods exist to help standardize the LCI phase and increase comparability between studies. The results of a study using different LCIA methods will vary based on differing classifications and CFs, but there generally exist agreements between the different methods (Dekker et al. 2019).

Whatever LCIA method is used must be clearly documented. A small selection of the available LCIA methods are described below.

8.3.1   TRACI

The Tool for the Reduction and Assessment of Chemical and Other Environmental Impacts (TRACI), developed by the US EPA, is the most common LCIA method used in North America. It was first developed in 2002 to bring a greater level of sophistication, comprehensiveness, and applicability to LCA studies in the United States (Bare 2012). Since then it has been updated, most recently to Version 2.1 in 2012. The TRACI methodology comes embedded in many LCA and LCIA tools, including those discussed in these chapters, or can be used standalone by download-ing the excel sheet from the EPA's website. TRACI is either the default or one of the LCIA options in every tool discussed in this chapter.

Currently, it features classifications and CFs for acidification, eutrophication, global climate change, ozone depletion, human health particulate, human health cancer, non-cancer, and ecotoxicity, photochemical smog formation, and fossil fuel resource depletion, with the intention to include land and water use in future updates (Bare 2012). These impact categories were chosen based on their international rec-ognition and relevance to US conditions. However, even just the selection of impact categories to include introduces bias to an LCA study. Each impact category has its own CFs. Some of them, such as global climate change and ozone depletion, are based on and sourced from internationally agreed upon practices, while others, including photochemical smog formation, were developed specifically for the United States (Bare 2012). Different impact categories carry with them various inherent levels of uncertainty.

8.3.2   ReCiPe

While TRACI is the most widely used impact assessment method in North America, it is not the only recognized one available. For studies outside of North America, other LCIA methodologies are more applicable. PRé Sustainability, the Dutch com-pany behind the SimaPro LCA software, in addition to RIVM, CML, Radboud Universiteit Nijmegen, and CE Delft, developed the ReCiPe impact assessment framework (Curran 2012). ReCiPe was initially launched in 2008 but updated in 2016 to provide globally applicable CFs, instead of just CFs relevant to Europe, as well as improve the existing characterization models (Dekker et al. 2019). The ReCiPe framework provides a large range of impact categories, including 18 mid-point indicators and 3 endpoint indicators (Huijbregts et al. 2016). It draws on the work of previous methods, including Eco-indicator 99 and CML 2001. Furthermore, there are choices in CFs for global and local conditions, as well as CFs that vary based on cultural perspectives. The LCIA method can be tailored toward an individualist, hierarchies, or egalitarian point of view (Huijbregts et al. 2016). The ability to tailor by location and cultural perspective, in addition to the broad range of impact categories included, makes ReCiPe one of the most diverse LCIA methodologies. Since this chapter is focused on North American conditions, the ReCiPe methodology is only available in SimaPro.

8.3.2.1 Eco-indicator 99

Before developing ReCiPe, PRé Sustainability had Eco-indicator 99, another LCIA tool and methodology. Unlike ReCiPe, Eco-indicator 99 focused on endpoint impacts (Monteiro and Freire 2012). Eco-indicator 99 takes a top-down approach and weights damages across three endpoint impact categories: human health, ecosystem quality, and resource use. These results can then be further combined down into one single score (Goedkoop and Spriensma 2000). Much like ReCiPe, Eco-indicator 99 includes different CFs based on the three cultural perspectives.

8.3.3 CML-IA

Also a predecessor to ReCiPe, the CML methodology, developed by the Center for Environmental Sciences at the University of Leiden, is a widely used set of indicators for studies outside of the United States (Dassault). At the same time, it is not commonly used in regard to studies concerning buildings (Monteiro and Freire 2012). It is based on European conditions and only features early-stage, midpoint impacts to limit uncertainty. Originally labeled CML 2001, the CFs were most recently updated in 2016 to the CML-IA (CML 2016). CML is one of the earliest LCIA methods originating in the early 1990s, and it was the first midpoint-oriented method (Curran 2012). It was extremely influential in shaping all subsequent LCIA methods. Of the tools discussed in this chapter, CML is only an option in SolidWorks Sustainability and SimaPro.

8.3.4 IMPACT World+

Another LCIA method that combines midpoint and endpoint impacts is the IMPACT World+ methodology, an update to the previously published IMPACT 2002+ method (Curran 2012). The IMPact Assessment of Chemical Toxicants (IMPACT) 2002+ method borrowed and adapted some midpoint indicators from existing methods, including Eco-indicator 99 and CML (Jolliet et al. 2003). The goal of IMPACT World+ was not only to update the classifications and characterization factors used in the methodology but also to offer regionalized methodologies at a global scale

(Curran 2012). The update recognizes and incorporates many advancements in LCIA methodologies, as well as address the need for consistent characterization factors for regionalized impacts (Bulle et al. 2019). Of the tools presented here, IMPACT World+ is only an option in SimaPro.

8.4 WBLCA Tools

Several LCIA tools exist for designers to use as a means of evaluating their work and making more informed design decisions. Tally, One Click LCA, Athena Impact Estimator for Buildings, and Solidworks Sustainability are all examples of tools that make the elaborate and rigorous LCA process more accessible. Designers are not LCA experts, and with the power of these tools they do not have to be. The dashboard of each of these programs acts as an interface between typical design workflows and LCA data. Workflows that traditionally have focused on describing, quantifying, and evaluating the aesthetic, performative, and financial project details can now integrate environmental impact data supported by scientifically supported LCA studies. Each program behaves differently but helps to answer the same LCIA questions. This chapter will take a detailed look at the dashboard of all four programs. Chapter 9 will demonstrate, through the lens of case study projects they were applied to, the situations each program is best suited to in addition to how to use them as decision-making tools for better, healthier designs.

These are not the only tools currently available, but they are four common tools in North America. Three of these tools—Athena, Tally, and One Click LCA—were designed with typical WBLCA practices in mind, while Solidworks Sustainability is typically used to assess the environmental impacts of smaller, more detailed design systems including building components.

8.4.1 A Note on Typical WBLCA Practices

The term "whole building" can be misleading when talking about these studies because they are not typically inclusive of every building element. Standard practice in North America is to include the building's envelope, structure, and floors (Bowick et al. 2017). These building elements typically account for over 90 percent of a building's total mass. Detailed interior assemblies, finishes, MEP systems, and site work are typically excluded from these assessments. Assessing environmental impacts on heterogenious lower mass elements increases model complexity, time, and cost without materially affecting the final assessment of the main stucture, core and shell of a building. Site work varies in area and volume and can easily be considered to fall outside the physical and methodological boundaries of a WBLCA. The object of assessment of each study should, however, be individually determined in tandem with the goal of each project and study.

8.4.1.1 Low Mass Material Assessment

One possible future area of research lies in developing specialized LCA/LCIA tools for designers not responsible for specifying and designing higher mass building elements. Interiors typically make up less than 10 percent of a building's mass. However, many of the elements that go into final fit out will be replaced several times during the functional life span of a building. In 100 years, a commercial interior can be expected to be completely replaced well over five times. If its mass is 5 percent of a building, it is not difficult to see how this becomes a material contributor to potential negative embodied environmental impacts.

It is important to note that many interior finishes companies are taking the lead one material at a time. Perhaps the best in class award goes to the Interface carpet company who uses robust LCA activities to guide what they produce and sell. At the time of this writing, Tally is beta-testing an interiors LCA functional unit module simulator. The examples in this book focus on larger scale designs because of the relative maturity of the interfaces to aid in environmental impact assessments at the building scale. The general approach can, however, be applied to smaller scale design project types as well an used by designers in numerous sustainable design disciplines.

8.4.2 Athena Impact Estimator for Buildings

The Athena Impact Estimator tool is one of the three LCA programs created by the Athena Sustainable Materials Institute, in collaboration with Morrison Hershfield, and is the only WBLCA program that they offer (Athena Sustainable Materials Institute n.d.). The Impact Estimator is a standalone tool that does not interface with any other software. Instead, users enter tabular information about building components, assemblies, and their specific quantities, and Athena generates a cradle-to-grave life cycle assessment about the building's impact. It operates with a proprietary database of North American LCI data that the Institute maintains (Yazdanbakhsh et al. 2018). Beyond their software, the Athena Institute, a nonprofit research collaborative, offers EPD services and advocates for LCA in the construction industry (Athena Sustainable Materials Institute n.d.). All of their programs are free to use. The information presented here is based on Version 5.4.0101. For more information, please visit www.athenasmi.org.

8.4.2.1 Establish a New Project

All functionality in the Athena dashboard can be found in the toolbar across the top of the screen and the project tree down the left side of the screen. The tool bar holds functionality regarding file management and reports, while the project tree holds functionality regarding the projects and material assemblies that make them up.

Fig. 8.1 Establishing a new project in Athena

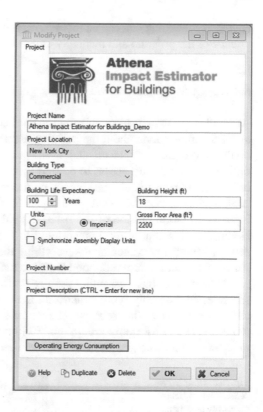

After right-clicking in the project tree and selecting the "New Project" option, a dialogue box will pop up asking for general information about the project (Fig. 8.1). This includes the general building location, building type, and approximate building size. The options for the building location represent major cities in North America, and the options for building type represent the major building typologies. The location chosen from this list informs Athena of general travel distances, material availability, and electricity grid mix.

When establishing a new project, there is a field to input the "Building Life Expectancy," which is a very important part of the study as it tells Athena how long the building is expected to be standing. The service life then determines how many times materials get replaced in the project based on their own expected life span, which is embedded in the LCI data. The closer this value is to the building's actual life span, the more accurate the results of the study will be, especially when balancing concerns of operational performance versus embodied impacts (Aktas and Bilec 2012). This service life is very important in understanding the results of a study and must be clearly reported. It will be determined in the goal and scope phase of the study. If two studies are not carried out across the same service life, the results are not comparable. Lastly, there is a button that says "Operating Energy Consumption." This brings the user to a pop-up window with the ability to enter the project's annual electricity, natural gas, LPG, heavy fuel, diesel, and gasoline use. Operational energy is something that can be included in a WBLCA study but must be calculated outside of the program and manually entered.

8.4.2.2   Add Project Assemblies

After all of the basic project information has been entered, the next steps are to enter all of the information about the building assemblies and quantities. Right-clicking on the project that was just created, in the project tree, brings up a menu of different options, including the option to add an assembly. Athena can calculate life cycle information about foundations, walls, columns and beams, roofs, and floors. There is also an assembly group called "Project Extra Materials" that allows the user to input information outside of the listed assembly groups. All of the materials and assemblies that can be used are limited to the available LCI data. The assembly groups offered above encompass everything that is typically included in a WBLCA study and cover the most common building material assemblies of the region.

Under each of these assembly groups, there are different options, appropriate to the selected assembly type, which represent common building constructions and can be used to enter the information about the project. For example, under the columns and beams assembly group, the "Columns and Beams" option represents the most common structural systems and is used in this example (Fig. 8.2). After

Fig. 8.2 Defining a "Column and Beam" assembly in Athena

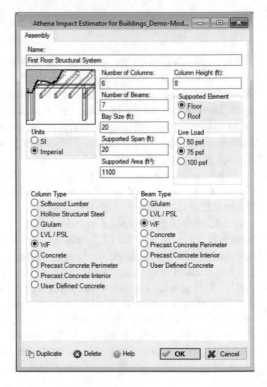

choosing it, a new window pops up to define the size and properties of the structural grid, starting with the number of columns, column height, and number of beams. This information is for only one floor of the project, and therefore the height of the columns may not match the full building height depending on the overall number of stories.

After adding in the basic information about the structural system, Athena asks for the bay size, supported span, and supported area. Then, the user must give information about the material of the columns and beams from a preset list of choices, including different steel, concrete, and wood options. Athena also asks for information regarding what the structural system is supporting, either a floor or a roof, and the live load it supports. This information will be used to calculate the sizing of the structural members. Throughout all building assemblies, Athena uses rules of thumb to size different members. The user cannot override these decisions.

The process of adding assemblies in all other groups is customized to the building component being defined, but the same approach is used to define the assembly in which the user inputs information about the assembly dimensions and materials. For example, when adding a wall assembly to a project, the "Custom Walls" option opens a window resembling that used to define the structural system (Fig. 8.3). First the user must add the length and height of the wall. Next, the user defines the primary assembly component (main structural layer) from a list of options that represent the typical wall constructions used in North America. This is only for the structural layer, and additional materials used in the wall assembly will be added later in the process. After selecting an assembly component, for example, steel studs, a new dialog box asks for information related to the sizing of that component. For steel studs, this information includes sheathing material, stud sizing, and stud spacing. More than one assembly component can be added to a single wall assembly if there is more than one structural layer.

After entering all of the basic information about the wall, there are two other sets of information to add relating to any openings in the wall and the envelope

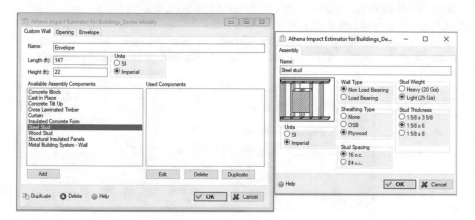

Fig. 8.3 Defining a new steel stud wall assembly in Athena

materials. Under the "Openings" tab, Athena asks for information on any windows and doors included in the wall being defined. Athena needs information regarding the number and size of openings as well as the materials that make up the windows or doors. Only one type of window or door can be added to a single wall assembly. The "Envelope" tab asks for information regarding the rest of the materials in the wall construction beyond the main assembly component. Any cladding materials, insulation, sheathing, or membrane layers are added here.

Adding any other assembly groups included in the study, including foundations, roofs, and floors, follows a similar structure with different options customized to that specific building component.

8.4.2.3 Additional Features

Athena has a few features built into the dashboard that make it easy for the user to save time and quickly understand what is happening as a result of the decisions they make. The first is the ability to copy an assembly. After one assembly has been added, right-clicking on it brings up a list of options, including the ability to copy it to any open projects. This can be useful to quickly add the same structural system on multiple floors, for example. By copying it to the same project, the user can then make any quick adjustments needed to the new assembly.

Athena provides users with the ability to see what materials, and in what quantities, are being used in the project. Clicking on the "Reports" tab in the toolbar and then "Bill of Materials" will allow the user to quickly see exactly what materials are being included in the WBLCA study (Fig. 8.4). This "Bill of Materials" is displayed by total quantity as well as the quantity split by building assembly group. This is a great way to understand exactly what is being included in a study as well as double check material quantities to verify the accuracy of the study. Items such as fasteners and adhesives that are a necessary part of the assembly and included in the study but not individually assigned by the user are displayed here. More information on the use of this information in a study can be found in Sect. 9.1.

Beyond generating a bill of materials from the assemblies defined within the program, Athena also provides users with the option to import a bill of materials from an excel, XML, CSV, or tab delimited text file, which can commonly be

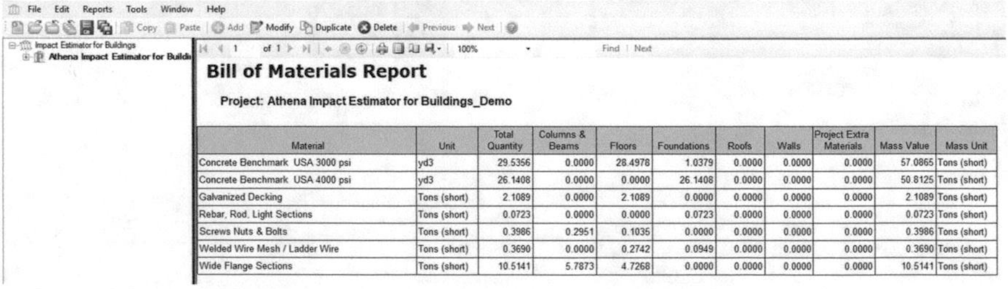

Bill of Materials Report

Project: Athena Impact Estimator for Buildings_Demo

Material	Unit	Total Quantity	Columns & Beams	Floors	Foundations	Roofs	Walls	Project Extra Materials	Mass Value	Mass Unit
Concrete Benchmark USA 3000 psi	yd3	29.5356	0.0000	28.4978	1.0379	0.0000	0.0000	0.0000	57.0865	Tons (short)
Concrete Benchmark USA 4000 psi	yd3	26.1408	0.0000	0.0000	26.1408	0.0000	0.0000	0.0000	50.8125	Tons (short)
Galvanized Decking	Tons (short)	2.1089	0.0000	2.1089	0.0000	0.0000	0.0000	0.0000	2.1089	Tons (short)
Rebar, Rod, Light Sections	Tons (short)	0.0723	0.0000	0.0000	0.0723	0.0000	0.0000	0.0000	0.0723	Tons (short)
Screws Nuts & Bolts	Tons (short)	0.3986	0.2951	0.1035	0.0000	0.0000	0.0000	0.0000	0.3986	Tons (short)
Welded Wire Mesh / Ladder Wire	Tons (short)	0.3690	0.0000	0.2742	0.0949	0.0000	0.0000	0.0000	0.3690	Tons (short)
Wide Flange Sections	Tons (short)	10.5141	5.7873	4.7268	0.0000	0.0000	0.0000	0.0000	10.5141	Tons (short)

Fig. 8.4 Bill of Materials generated from the Athena "Reports" tab

exported from different BIM and CAD programs. In the case of an excel file, after it is brought in, Athena looks for four basic parameters in the data: what material it is, what assembly group it belongs to, what quantity of material there is, and what unit this quantity is in. All of this information is then mapped to existing parameters in Athena and then added to the corresponding assembly group as additional materials, or just generically to the project. This can be a time-saving method to use Athena in tandem with Revit or other BIM software.

Athena provides users with the option to quickly reference the impacts of different assemblies in the project tree. The user can display the impacts of different assemblies as either a percentage or value right in the project tree for the selected environmental impact category. This useful feature allows a designer to quickly see the most impactful elements of their design, in one impact category, as they add information about their project.

8.4.2.4 View Results

While displaying the impacts in one category in the project tree is a quick way to identify and test targeted results of the study, the full results of the study can be accessed from the "Reports" tab. On this screen the user has the ability to view the WBLCA study results by life cycle stage or by assembly group (Fig. 8.5). The user also has the ability to define the system boundary, either A–C or A–D. These boundary categories reference the life cycle stages included in the study as defined in EN 15978, Sustainability of construction works—Assessment of environmental performance of buildings—Calculation method. Life cycle stages A1–A3 (product); A4

Condensed LCA Measure Table By Life Cycle Stages

Project: Athena Impact Estimator for Buildings_Demo

LCA Measures	Unit	PRODUCT (A1 to A3) Total	CONSTRUCTION PROCESS (A4 & A5) Total	USE (B2, B4 & B6) Replacement Total	USE Operational Energy Use Total	USE Total	END OF LIFE (C1 to C4) Total	BEYOND BUILDING LIFE (D) Total	TOTAL EFFECTS A to C	TOTAL EFFECTS A to D
Global Warming Potential	kg CO2 eq	4.23E+04	6.04E+03	1.09E+04	0.00E+00	1.09E+04	2.06E+03	-1.57E+04	6.12E+04	4.55E+04
Acidification Potential	kg SO2 eq	2.11E+02	6.12E+01	8.87E+01	0.00E+00	8.87E+01	2.40E+01	-8.99E+01	3.85E+02	2.95E+02
HH Particulate	kg PM2.5 eq	8.75E+01	4.68E+00	6.76E+00	0.00E+00	6.76E+00	1.96E+00	-9.43E+00	1.62E+02	1.52E+02
Eutrophication Potential	kg N eq	2.74E+01	4.82E+00	2.69E+00	0.00E+00	2.69E+00	1.49E+00	-1.57E+00	3.64E+01	3.48E+01
Ozone Depletion Potential	kg CFC-11 eq	5.76E-04	5.49E-05	1.96E-04	0.00E+00	1.96E-04	8.59E-08	-5.87E-07	8.27E-04	8.27E-04
Smog Potential	kg O3 eq	2.65E+03	1.80E+03	9.17E+02	0.00E+00	9.17E+02	7.78E+02	-7.05E+02	6.14E+03	5.44E+03
Total Primary Energy	MJ	5.76E+05	8.11E+04	3.50E+05	0.00E+00	3.50E+05	3.02E+04	-1.30E+05	1.04E+06	9.07E+05
Non-Renewable Energy	MJ	5.60E+05	7.99E+04	3.50E+05	0.00E+00	3.50E+05	3.02E+04	-1.29E+05	1.02E+06	8.90E+05
Fossil Fuel Consumption	MJ	4.66E+05	7.67E+04	3.47E+05	0.00E+00	3.47E+05	3.01E+04	-1.31E+05	9.20E+05	7.88E+05

LCA Measure Table By Assembly Groups (A to D)

Project: Athena Impact Estimator for Buildings_Demo

LCA Measures	Unit	Foundations	Walls	Columns and Beams	Roofs	Floors	Project Extra Materials	Total
Global Warming Potential	kg CO2 eq	8.76E+03	6.49E+03	6.33E+03	0.00E+00	2.39E+04	0.00E+00	4.55E+04
Acidification Potential	kg SO2 eq	3.33E+01	1.00E+02	4.00E+01	0.00E+00	1.00E+02	0.00E+00	2.95E+02
HH Particulate	kg PM2.5 eq	6.27E+00	1.86E+01	2.63E+01	0.00E+00	1.01E+02	0.00E+00	1.52E+02
Eutrophication Potential	kg N eq	9.34E+00	8.93E+00	1.09E+00	0.00E+00	1.53E+01	0.00E+00	3.48E+01
Ozone Depletion Potential	kg CFC-11 eq	1.89E-04	3.68E-04	1.23E-05	0.00E+00	2.57E-04	0.00E+00	8.27E-04
Smog Potential	kg O3 eq	7.45E+02	2.03E+03	4.73E+02	0.00E+00	2.19E+03	0.00E+00	5.44E+03
Total Primary Energy	MJ	6.31E+04	1.76E+05	1.12E+05	0.00E+00	5.56E+05	0.00E+00	9.07E+05
Non-Renewable Energy	MJ	6.10E+04	1.65E+05	1.12E+05	0.00E+00	5.53E+05	0.00E+00	8.90E+05
Fossil Fuel Consumption	MJ	5.70E+04	1.48E+05	6.89E+04	0.00E+00	5.15E+05	0.00E+00	7.88E+05

Fig. 8.5 LCIA results from Athena by life cycle stage and assembly group

and A5 (construction process); B2, B4, and B6 (use); C1–C4 (end of life) are included in the study. However, the user also has the ability to include stage D (beyond building life). Regardless of which decision is made, the boundary of the system must be clearly documented and transparently communicated along with the results of the study.

Clicking "Show Report" will bring up a new window with a table of results across the included environmental impact categories categorized by either life cycle stages or assembly groups depending on which option the user chose. The environmental impact categories included in the study are GWP (in kg CO_2eq), acidification potential (in kg SO_2eq), HH particulate (in kg PM2.5eq), eutrophication potential (in kg Neq), ozone depletion potential (in kg CFC-11eq), smog potential (in kg O_3eq), total primary energy (in MJ), nonrenewable energy (in MJ), and fossil fuel consumption (in MJ). Athena uses the TRACI LCIA methodology. The table of results can be exported to a PDF, word file, or excel sheet. Also in the "Reports" window is the ability to generate reports specific to two different green rating systems local to North America, Green Globes and LEED.

The Export submenu under the File menu provides the user with the ability to export a more detailed PDF report or excel report of the results, beyond those that can be generated from the table of results in Athena. The PDF report starts with information about the Athena Institute, the Impact Estimator for Buildings, and other information about the software and methodologies used in the study. After this information the PDF breaks down the results of the study in tabular format by life cycle stage and assembly group. Next, the results of the study are provided in both graphical and tabular format by the assembly group for each environmental impact category individually. Information about embodied carbon over the life cycle of the building per project area (in m^2) is given next. The next section of the report breaks down operational versus embodied impacts, in both graphical and tabular formats. Lastly, the report concludes with the bill of materials.

The excel report that is generated gives a more detailed breakdown of the information as well as the ability for the user to compare the results of multiple studies. The file reports the same general project information, bill of materials, and environmental impacts by life cycle stage. It also reports the energy consumption of the project across the different life cycle stages and broken down into over a dozen different energy sources. The same process is repeated for air emissions, water emissions, land emissions, and resource use. These flows of matter and energy represent the results of the LCI phase of the study and are further described in Chap. 7. Also embedded in the excel file is a utility that allows the user to compare the results across multiple studies. The "Report Difference" tab allows the user to generate a report that shows the difference in impact between two selected projects. These tools are useful in fully comparing two different design options and making the most informed decision. If two studies are carried out using Athena, and the user inputs the same information regarding goal and scope, then the two studies are fully comparable. Making comparisons between design options is one easy way to act on the results of the study. Examples of how Athena can be used to inform design decisions can be found in Chap. 9.

8.4.3 Tally

Tally is a WBLCA add-in for Autodesk Revit. It was developed by Philadelphia-based architecture firm KieranTimberlake in conjunction with Autodesk and Sphera (formerly thinkstep) (Cays 2017). Tally is designed to be used iteratively throughout the design process, from schematic design on. It offers either a full building study or design option comparison. By assigning a relationship between BIM materials and LCI data, which Tally sources from the German-based company Sphera, Tally is able to provide designers with graphs that break down the environmental impacts of a design across eight impact categories and five life cycle stages (KT Innovations n.d.). These impacts can be viewed in broad categories, by material division or Revit category or by individual products. Tally is a US-based application, and assessment results are driven primarily by US data sources. The information presented here is based on Version 2020.06.09.01. For more information, please visit www.choose-tally.com.

8.4.3.1 Establish a New Study

Tally operates as a plugin to Revit and can be accessed from the "Add-Ins" tab in the program. To start using Tally, the first decision that needs to be made is defining the object of study. The three options are "Full Building Study," "Design Option Comparison," and "Define Template File." "Full Building Study" is the appropriate object of study to complete a WBLCA. The next step, after choosing "Full Building Study," is to define which categories, work sets, and phases will be included in the study. This method of filtering allows users to match the goal and scope of the project without having to alter the Revit model.

8.4.3.2 Assign Materials

After defining what will be included in the study, the next step is to match the material information from the Revit model to Tally's LCI database. All of the Revit materials are listed out, organized first by category, then Revit family, and then the individual materials that make up the families. To assign data to a material, right-click on the material, and a new menu with options, including the ability to define the material, will open.

After picking the appropriate Tally entry to match to the Revit material, a new window pops up to define more specific information about it. In the case of a structural steel column, there are choices for what type of steel is being used. The next category, service life, is important in defining how long the material will be in the building before getting replaced. The building's structure will not be replaced throughout the duration of the project, but something like gypsum board would be replaced more frequently, and this is where that parameter would be set.

After setting the service life, the next category asks for the material's takeoff method. This is how Tally interprets the information from Revit to know how much of a material is being used in the project. For a steel column, this can be done by the volume modeled in the Revit file or by the linear length of column (present in the file) and the section of the column, as defined by the user in Tally. Understanding how the Revit model was built will help the user to pick the most appropriate takeoff method. These options vary based on the type of building element being defined.

These initial components of the LCI definition (material, service life, and takeoff) are common across all Tally entries. After these are defined, Tally presents additional options relevant to the specific building element being defined, if there are any relevant options (Fig. 8.6). In the case of the steel column, options for fireproofing as well as different finishes are provided to the user. This allows for a more holistic study that best reflects the real-world conditions of how the building will be built. If any additional materials are defined, then the user must set the service life and takeoff method for the individual additional component. The importance of additional materials to the accurate results of a study can be found in Sect. 9.1.1.

Something like a column is modeled in Revit very closely to how it would exist in the real world, but this is not the case for all building elements. For instance, a metal stud layer in a wall construction is modeled in Revit as a solid rectangular extrusion. This is not how it is built in reality. For this reason, takeoff for metal studs is calculated based on their spacing (Fig. 8.7). The user inputs the stud section and

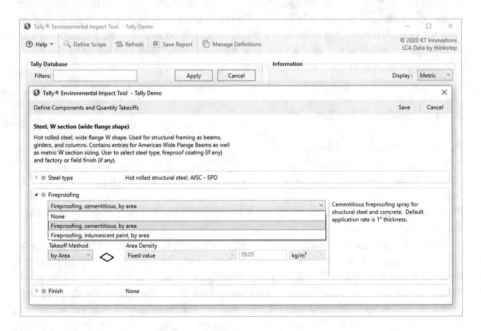

Fig. 8.6 Additional material options for fireproofing and finish presented when defining a steel column in Tally

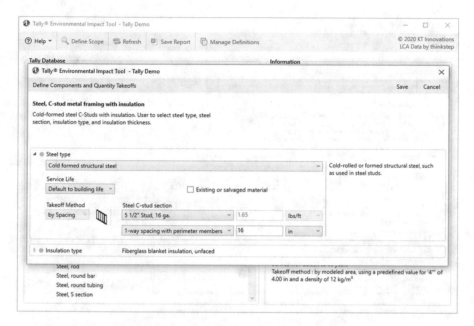

Fig. 8.7 Takeoff method by spacing for metal stud layer

the spacing between the studs. Tally uses this information to calculate how much of the material will be used based on the area of the wall.

Another notable area that Revit models differ from real-world conditions is the lack of smaller details within a family. For instance, window hardware is not modeled in a Revit window family. Tally recognizes this and when specifying a window frame, the option to include window hardware is also included to reflect real-world conditions and provide more accurate results of the study (Fig. 8.8). These choices are presented as the additional materials in the Tally entry. While each Tally entry is tailored to the specific Revit material and building component being assigned, the overall process remains the same for materials in every category and family.

8.4.3.3 Additional Features

Beyond defining materials, there are several additional features and functionalities built into the Tally dashboard to aid users. After adding a material definition from Tally to a Revit material, the box on the right of the dashboard is updated to reflect what information was defined so that all assumptions are clearly laid out (Fig. 8.9). It quickly outlines the overall description of the chosen Tally entry, the quantity of the material included in the study, the service life, takeoff method, and any additional materials added. Transparency in LCA studies is very important, and this is one way that Tally works to extend it into the design workflow. Another feature of Tally is the ability to copy material definitions between families in the same

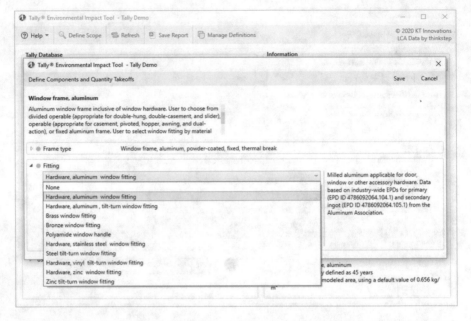

Fig. 8.8 Window fitting options included in Tally entry to compensate for the lack of any hardware or fittings included in the Revit model

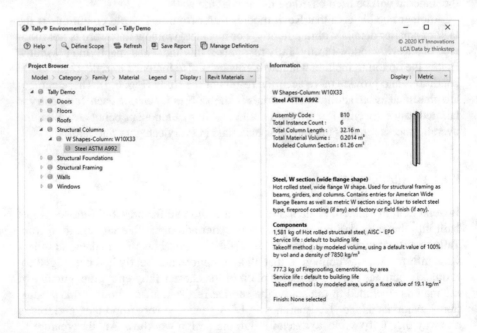

Fig. 8.9 Information panel on the right of the dashboard with updated assumptions based on the Tally material entry

category. This is a time-saving measure when the same material is used across families. For instance, if multiple window sizes have the same frame or glass material, the definition can easily be copied between them. The "Isolate in Revit View" option allows users to quickly see where in the Revit model a material is located to better understand what they are defining. Any information entered in Tally is automatically saved so long as the Revit file itself is saved. This gives users the ability to come back to the WBLCA study as they are working, to update any material definitions or to rerun the study after they have made changes to the design in Revit.

8.4.3.4 Save Report

Once all of the materials have been assigned in Tally, hitting the "Save Report" button will lead to the last round of information that must be entered before generating the results of the study. This page asks for general information about the project including the location and gross building area. The "Expected Building Life," or service life, is determined in the goal and scope phase of the study and is entered here. Tally then provides a box for the user to write out the goal and scope of the assessment. Carefully documenting this is important in using and understanding the results of a study, especially when trying to interpret the results of a study someone else carried out.

Tally defaults to including biogenic carbon, but this is an option that can be controlled by the user. Biogenic carbon is a widely debated topic surrounding LCA studies. Organic materials, such as wood, sequester carbon as they grow. Although processing wood to be used in a building will release some carbon dioxide, this is often far less than the amount sequestered in the material itself. According to a study conducted by Arup on the global warming/cooling effects of timber structural materials, the emissions account for less than half of the sequestered carbon, even with longer transport distances (Fig. 8.10). This is only true of wood that comes from a sustainably managed forest, however. Timber from non-sustainable forestry practices still has the potential to emit less carbon than other structural alternatives, like concrete, but overall the wood structure still has a "positive" (meaning carbon dioxide will be released into the atmosphere not that the overall results are beneficial) carbon footprint.

Including biogenic carbon in a Tally study will account for a "negative" carbon footprint for these organic materials, meaning they store more carbon than is released in the manufacturing process. However, there are several approaches to accounting for biogenic carbon. Each has its own challenges. Issues ranging from increased uncertainty to not accounting for deforestation make it a difficult topic to handle within an LCA study (Carbon Leadership Forum 2020). One argument in favor of considering biogenic carbon in the calculation, especially when wood makes up a large percentage of the overall mass of a building, recognizes the useful life span of a building and the benefit conferred by sequestering carbon over many decades and possibly a century or more. This appropriately intervenes over the time scale to mitigate the negative environmental impacts of an excess buildup of GHG

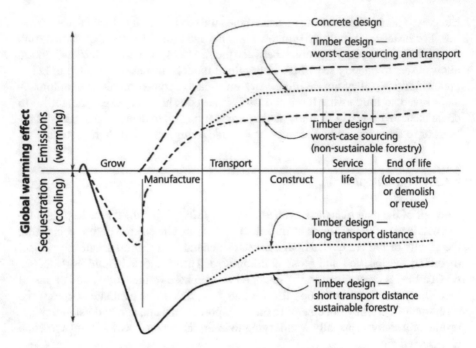

Fig. 8.10 Comparison of carbon emissions between timber and concrete structural materials for a 12-story tower, with scenarios for sustainable and non-sustainable forestry conditions. (Credit Bruce King, based on work of Arup engineers (King 2017). Used with permission)

in the atmosphere. Biogenic carbon is not an option in either of the other WBLCA tools discussed in this chapter but is the default setting in Tally. Whether or not it is included, it must be clearly stated.

By default Tally runs a cradle-to-grave LCA study inclusive of EN 15978 life cycle stages A1–A3 (product), A4–A5 (construction stage), B2–B6 (use stage), C2–C4 (end of life), and module D (reuse and recovery). To account for A4 (transportation), Tally includes default travel distances for materials being brought to the site (Fig. 8.11). They all assume the materials are being brought to the site via truck and include different travel distances for different materials. These estimates are not specific to the building location entered. They serve as good estimates for an early-stage WBLCA study, but as more information is known about where products are being sourced from, how far they will be traveling, and what vehicle they will be transported on, these numbers should be updated accordingly. Tally also gives the user the option to include impacts from stages A5 (construction installation) and B6 (operational energy). These include electricity, heating, and water use based on regional grids across the United States and globally. Tally does not provide any guidance on these measures, and they must be calculated separately and input manually.

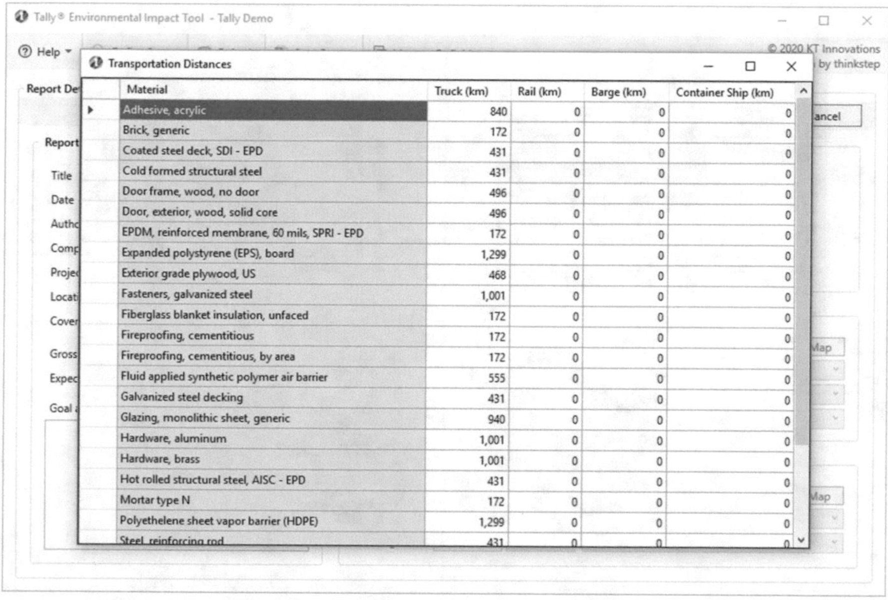

Material	Truck (km)	Rail (km)	Barge (km)	Container Ship (km)
Adhesive, acrylic	840	0	0	0
Brick, generic	172	0	0	0
Coated steel deck, SDI - EPD	431	0	0	0
Cold formed structural steel	431	0	0	0
Door frame, wood, no door	496	0	0	0
Door, exterior, wood, solid core	496	0	0	0
EPDM, reinforced membrane, 60 mils, SPRI - EPD	172	0	0	0
Expanded polystyrene (EPS), board	1,299	0	0	0
Exterior grade plywood, US	468	0	0	0
Fasteners, galvanized steel	1,001	0	0	0
Fiberglass blanket insulation, unfaced	172	0	0	0
Fireproofing, cementitious	172	0	0	0
Fireproofing, cementitious, by area	172	0	0	0
Fluid applied synthetic polymer air barrier	555	0	0	0
Galvanized steel decking	431	0	0	0
Glazing, monolithic sheet, generic	940	0	0	0
Hardware, aluminum	1,001	0	0	0
Hardware, brass	1,001	0	0	0
Hot rolled structural steel, AISC - EPD	431	0	0	0
Mortar type N	172	0	0	0
Polyethelene sheet vapor barrier (HDPE)	1,299	0	0	0
Steel reinforcing rod	431	0	0	0

Fig. 8.11 Default transport distances included in Tally. These options can be updated by the user to more closely reflect actual project conditions if they are known

8.4.3.5 View Results

Tally takes all of this information and combines it into a final PDF report of LCIA results. This shows an overview of the study and results. The "Report Summary" quickly outlines key information and assumptions about the study. Then, the results of the study are broken down by impact category and life cycle stage. The impact categories included in Tally are GWP (in kg CO_2eq), acidification potential (in kg SO_2eq), eutrophication potential (in kg Neq), smog formation potential (in kg O_3eq), ozone depletion potential (in kg CFC-11eq), Primary energy (in MJ), nonrenewable energy (in MJ), and renewable energy (in MJ). Of these impact categories, the impacts of GWP, acidification potential, eutrophication potential, smog formation potential, and nonrenewable energy are prioritized and broken down graphically by life cycle stage, division, Tally entry, Revit category, and Revit family (Fig. 8.12).

All of these different ways of breaking down the results are helpful in identifying any hotspots where the largest environmental impacts are found to help the designer focus their attention on the most impactful areas of the project, as seen in Sect. 9.1.2. The rest of the report generated by Tally provides critical information about the methodologies and assumptions of the study, most of which are pre-established by Tally and out of the hands of the designer. All of this information is crucial to revealing the parts of the process that happen without user input. Along with this final report, Tally also generates an excel spreadsheet that gives a more detailed breakdown of all of the impact categories across the same life cycle stages,

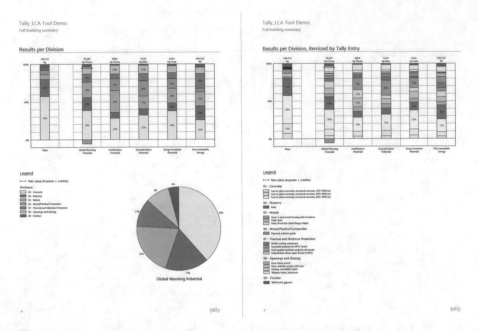

Fig. 8.12 Tally report showing results per division and itemized by Tally entry. Results for five impact categories are displayed graphically with global warming potential being highlighted

divisions, Tally entries, Revit families, and materials. This information is very useful in further understanding the results and finding targeted ways to lessen the overall impacts of the design. Examples of how Tally can be used to inform design decisions can be found in Chap. 9.

8.4.4   One Click LCA

Operating at the same scale of a whole building, One Click LCA is an LCIA tool that can operate standalone or interface with Revit, and other BIM software, to analyze the impacts of a building's design. However, One Click LCA goes beyond LCA studies and offers designers the ability to perform life cycle costing, circularity assessment, carbon benchmarking, and more. One Click LCA also provides information on compliance with several worldwide green design certifications, including LEED v.4 and Living Building Challenge (Bionova n.d.). While Bionova, the company that developed One Click LCA, is based in Finland, the LCI data is made up of many databases and is representative of global conditions. One Click LCA sorts this information into geographical databases, offering information specific to North America, Europe, South America, Asia Pacific, and the Middle East (Bionova n.d.). The information presented here is based on a North American business license in July 2020. For more information, please visit www.oneclicklca.com.

8.4.4.1   Define Object of Study

One Click LCA functions both as a web-based LCIA program and as a Revit Plugin. While the program supports studies in over 60 countries, this will focus specifically on the applications of the tool using a North American license. Furthermore, the software integrates with more BIM programs than just Revit, but again this will only focus on the Revit integration. The Revit interface is accessed from the "Add-Ins" tab. Options for both "LCA in Cloud" and "LCA in Revit" will be available, but this description will focus on the "LCA in Revit" option. The first page displayed after launching One Click LCA is a page to select what categories will and will not be included in the study. In addition to choosing what categories the study is inclusive of, One Click LCA also provides options for the default unit to be used in takeoff calculations. The default setting is volume, which uses the three-dimensional geometry available from the Revit model, but area and number of units are also both options. This setting can be changed later on in the process for individual materials.

8.4.4.2   Assign Materials

After defining the scope of the study, LCI data from One Click LCA's extensive database is matched to material entries from Revit. This can be done manually, as described here, or automatically. The automatic mapping provides an additional time-saving method, and the user has the ability to update any of the assigned mappings to match the specifics of the project. All of the Revit materials are organized only by category, not separated out by the Revit family. One Click LCA offers several different LCI databases sorted by geographical location, explained in Chap. 7. This book is rooted in North American practice, and therefore North American data is being used. After choosing a database the user has the option to further filter the available LCI data before matching it to the Revit entries. The first filter parameter gives the user the ability to filter by a wide range of material categories (Fig. 8.13). If a material category is selected, then only materials included in that category appear when the user starts to type in the box to map the Revit entry to the One Click LCA database. Beyond filtering by material category, LCI data entries can also be sorted by country and type. The last LCI data filter allows the user to filter by data type. The three options are private, generic, and manufacturer. The type of data to be used in a study should always be determined during the goal and scope phase.

Once the appropriate filters have been set, material information can be defined. After clicking in the box "Map to One Click LCA," a list of relevant LCI data points (based on the filters) are displayed. For each data entry, the name of the material, material properties (thickness and/or density), company (if the data is specific to a manufacturer), country of origin, and source are all displayed as one-line item (Fig. 8.14). This allows users to quickly evaluate all options for the material in question and pick the most relevant data source. If no filters are set, all of the options will

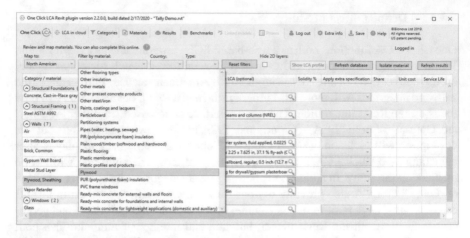

Fig. 8.13 Material categories to filter the available LCI data

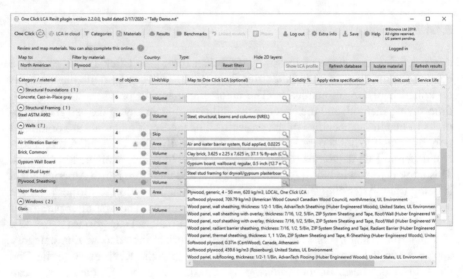

Fig. 8.14 LCI line entries showing information on where and how the dataset was sourced as it is mapped to the Revit materials

appear and then slowly filter down as the user types what material they are looking for.

After defining the material, there are more options related to how it is calculated, including takeoff methods. The options for how the material quantity is counted are volume, area, or skip. Volume is the default unit and will be the most accurate for most cases. To the left of the LCI data point, there is an option to indicate "Solidity %." If this field is left blank, the default solidity is 100%, meaning that the modeled volume is fully solid. However, the user can override this by entering a volume. In the case of a material that is modeled in Revit very different from how it is built in real life, the solidity is often accounted for in the LCI entry, and this box can be left blank. For example, in the case of a metal stud layer in a wall, the LCI entries specify the stud size and spacing and use the dimensions of the wall in Revit to get material takeoffs. One of the LCI data entries available to users in this case is "Steel stud framing for drywall/gypsum plasterboard per sq. meter of wall area (incl. air gaps per m³), C-profile: 4×2 inch, gauge 25, 10 ft. height $\times$ 16 inch (40 cm) spacing." This dataset accounts for the differences in how metal studs are modeled and actually built.

The "Apply extra specification" field can be used to add rebar to any concrete materials, and the share field defines how much rebar is present. The "Unit cost" field is used for One Click LCA's life cycle costing tool, which is separate to the WBLCA. Lastly, the service life of the individual material being defined can be entered here if it is known. Otherwise, if this field is left blank, a default service life based on the LCI entry chosen will be applied to the material. The user can later see or change this service life in the web interface. The process of assigning materials is the same across all material categories. After all of the Revit materials have been defined, the "Save" button will store the LCA data with the Revit model.

8.4.4.3 Additional Features

One Click LCA has data transparency embedded in both the Revit and web interface. After selecting an LCI data point, clicking "Show LCA Profile" will bring up metadata on the dataset. For more information on the importance of data transparency, see Chap. 7. In the Revit interface, several built-in features help the user to better understand what information from the Revit model One Click LCA is translating into the study. Next to the Revit material name is a "# of objects" count to tell you how many times that specific material appears in the file. Clicking on the question mark icon next to this pops open a window with the material thickness and total area. This can be used as a sense check to make sure that the model is accurate. Using the "Isolate Material" function will temporarily hide all of the Revit model with the exception of the family that the material in question belongs to.

8.4.4.4 Establish a New Study

The rest of the WBLCA process, including entering general project information as well as generating results, happens in the One Click LCA web interface. Before bringing the results to the web, a project must be created to host the study. In the web interface, studies are organized by project and design. The project holds overarching information regarding the study and can host many different design options of the same project. The information required in a project includes the project name, address, gross floor area, number of stories above grade, building type, and frame type. One Click LCA has built-in functionality to support a large number of green building certifications from around the world, and the certification being pursued, if any, can be designated when setting up a new project.

Once the project has been created in the web interface, the "LCA in cloud" button in the Revit interface will bring the material quantities and LCI data mappings from Revit to the web to complete the study. The study must get entered as a design within the project that was created. The ability to host multiple designs within one project allows for easy comparisons of the environmental impacts of different designs. Information specific to the design must be entered, including a unique name for the design, what stage of the design process the study is being completed at, what type of project it is, the frame type, and what overarching building component types are included in the study. All of this information will have been previously defined in the goal and scope phase of the study and reported here.

8.4.4.5 Additional Information Required

As the data is moved to the web, the user has the ability to again visit the material mappings and quantities included in the study as defined in Revit. The materials from the LCI data points that were assigned are displayed, along with their quantity, the assumed transportation method and distance, and the service life of the specific product (Fig. 8.15). The transportation and service life are set using defaults based on the product and location, but the user has the ability to update the information if data specific to the project is known.

Fig. 8.15 LCI material mappings, quantities, transport distances, and service lives in the One Click LCA web interface

One Click LCA requires the user to input information on annual energy consumption, calculation period, and building area, though results can still be generated without them. When inputting the annual operational energy, the user is asked for the total electricity use in kWh along with the electricity grid profile being used in the project, based on the project location. The calculation period is the expected service life of the building and will be determined in the goal and scope phase of the study. The building area is used to benchmark the results by square foot or square meter.

8.4.4.6 View Results

After all of this information has been entered, the results of the study can be viewed right in the web interface. The impact categories included vary based on the compliance requirements that govern the tool being used. In this case the "LCA compliant with US codes and sustainable building standards" provides impacts in GWP (in kg CO_2eq), acidification (in kg SO_2eq), eutrophication (in kg Neq), ozone depletion (in kg CFC-11eq), formation of tropospheric ozone (in kg O_3eq), fossil fuel primary energy (in MJ), and total use of primary energy (in MJ). This specific study is cradle-to-grave inclusive of EN 15978 life cycle stages A1–A3 (construction materials), A4 (transportation to site), B1–B5 (maintenance and material replacement), B6 (energy use), and C1–C4 (deconstruction).

The results are first displayed in tabular form by life cycle stage and impact category (Fig. 8.16). Further down on the results page is a section called "Most

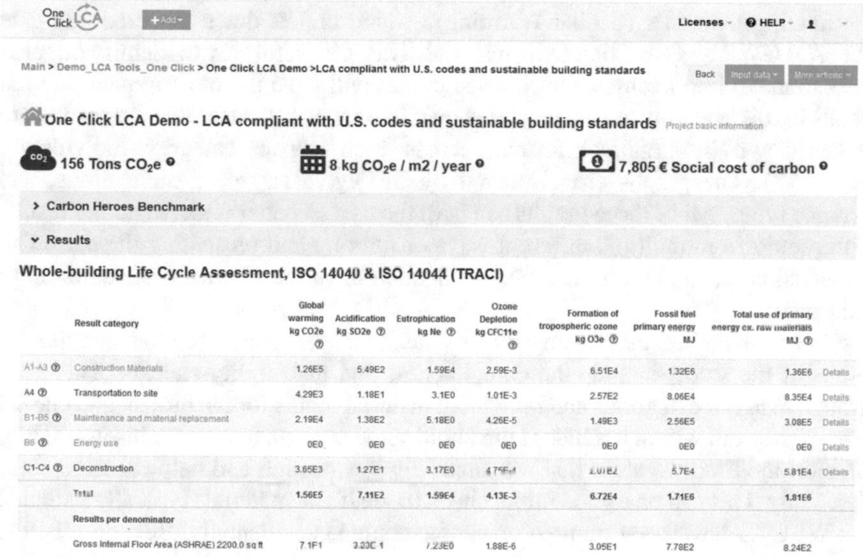

Fig. 8.16 Tabular results of LCA study by life cycle stage and impact category

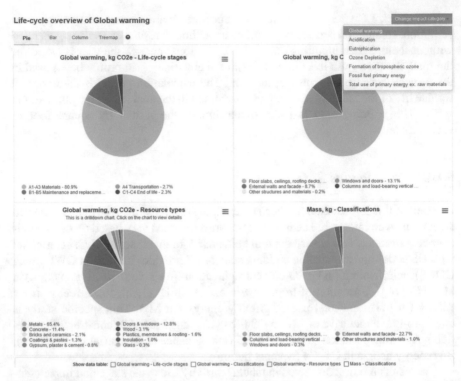

Fig. 8.17 Graphical breakdown of global warming potential in the project. These same graphs can be displayed for any impact category included in the study

contributing materials (Global Warming)" which breaks down the cradle-to-gate (A1–A3) GWP by One Click LCA material. This is a useful way to identify targeted areas to make improvements in the design that will have the most impact on total results in this one impact category as shown in Sect. 9.1.2. After this the results are broken down in graphical format, across each impact category individually (Fig. 8.17). The graphs are displayed by life cycle stages, classifications, and resource types. All of these graphs can help the user to better understand the results of the study from multiple different ways of filtering and presenting the data. The individual charts and tables can be downloaded in various formats, including PDF and excel.

Some of the other features on the One Click LCA dashboard, beyond the reported results of the study, include the completeness and plausibility checker. This goes through material categories and individual material inputs for the project and checks to see if they fall within a standard threshold value per square meter. This useful tool can alert the user of values that are unusually low or high and help ensure a complete study. This can be a good initial check to see if any materials typically included in a WBLCA have been omitted or double counted accidentally. Beyond this, the

Carbon Heroes Benchmark provides a comparison of the embodied carbon of the study, per square meter, against other projects with similar parameters. However, benchmarking WBLCA studies is very difficult to do based on the number of parameters that can affect the results of the study (Simonen et al. 2017). Examples of how One Click LCA can be used to inform design decisions can be found in Chap. 9.

8.5 LCIA Tools at Different Scales

While the scale of a whole building, or even a city, is at one extreme when talking about the object of assessment for an LCA study, there are many other scales at which the results of the study prove beneficial. A building is made up of a series of different systems and components. An LCA study can be carried out on any of the discrete building parts, products, or furnishings. While upfront it seems like a study at this scale has much less potential to mitigate impacts across the life cycle of the building, it is important to remember that any savings on the level of an individual product gets multiplied up to the scale of the overall building based on the number of units used in a building design. Furthermore, individual elements of a building get replaced multiple times throughout the life span of a building. Specifically, interiors and furnishings are updated frequently, often before they have outlived their potential (Aktas and Bilec 2012). This makes savings at the level of an individual product very advantageous throughout the full life cycle of a building. Additionally, optimizing environmental impact reduction on a specific building product that is manufactured, marketed, and sold on an industrial scale does not benefit the impact profile of only a single building.

8.5.1 SolidWorks Sustainability

Another useful LCIA tool for designers is SolidWorks Sustainability, an LCIA tool integrated into SolidWorks. SolidWorks is a program developed by Dassault Systèmes primarily for engineers and product designers (Dassault Systèmes n.d.). SolidWorks Sustainability is very useful for looking at the life cycle impacts of an individual product or components and all of the pieces that make it up. The dashboard indicates percent change of impacts between design changes across four impact categories. Like Tally, LCI data is also sourced from Sphera (Dassault Systèmes n.d.). For more information, please visit www.solidworks.com/sustainability/.

8.5.1.1 Access the Dashboard

SolidWorks has two different sustainability products. SolidWorks SustainabilityXpress comes with the software and has limited functionality. SolidWorks Sustainability comes with SolidWorks Premium and has expanded functionality. This chapter is presenting the functionality and usage of the full SolidWorks Sustainability. SolidWorks Sustainability is accessed from the "Evaluate Tab" within SolidWorks. Opening it brings up a task pane on the right of the screen with all of the inputs, as well as results clearly laid out for the user. Information is entered, and results are generated real time, for both part and assembly files which gives the user the ability to critically analyze and improve their design at multiple stages and scales throughout the design process. The results of the study are displayed for the life cycle stages material, manufacturing, use, end of life, and transportation. Impacts are displayed for carbon footprint (in CO_2eq), energy consumption (in MJ), air acidification (in SO_2eq), and water eutrophication (in PO_4eq). The dashboard also provides real-time updates on the financial impacts of the design, with a simplified LCC built in.

8.5.1.2 Part File: Assign Materials

In a part file, the user is asked to define information on the material, manufacturing, use, transportation, and end of life. Starting with the material that the part is made of, the options are organized first by class and then name. There are many different, highly specific material choices to pick from for each material class. After defining the material, a default recycled content based on default trade mixes is displayed and can be updated. Underneath this, the weight (in pounds) of the part is displayed.

After defining the material, the user then needs to give information on the manufacturing process of the part, starting with the region where the part is manufactured and how long it is manufactured to last. SolidWorks Sustainability has data available for North America, Europe, Asia, Japan, South America, Australia, and India. This information is used for the manufacturing technologies and energy used in creation. Lastly, the user enters information about the type of manufacturing process being used and the finish or coating on the part, if any. The user can also input information about the total energy usage, total natural gas usage, and scrap rate of the manufacturing process.

The remaining information required relates to the use, transportation, and end of life of the product. The "Use" section asks the user to select the region the part will be used in. In the "Transportation" section, the user can see the default transportation methods and distances of the part from manufacturing to use as well as update them to be specific to the project. Lastly, the "End of Life" section asks the user for information about how the part will be disposed of at the end of its life based on percent recycled, incinerated, and sent to landfill. Default values, based on the selected material, are set, but the user can enter more specific information if it is available to them.

8.5.1.3   Additional Features

Within the "Material" section is a button labeled "Find Similar" which helps users to reduce the environmental impacts of the part based on the material choice. The user can choose to set criteria for a variety of material properties that the material must meet. These properties include elastic modulus (in N/m^2), shear modulus (in N/m^2), thermal expansion coefficient (per K), density (in kg/m^3), and tensile strength (in N/m^2). The conditions for each material property include any, which signifies that the value of the property is not important, or <,>,~ a value. After setting the criteria, pressing the "Find Similar" button will bring up a list of all of the materials that satisfy the requirements and their material properties. Clicking on a material will show the percent change across the four impact categories, as well as financial impacts. The "Manufacturing Process" input gives the user the ability to select the type of manufacturing that will be used, and the environmental impacts are updated accordingly.

8.5.1.4   View Results

Beneath all of the panels to enter information is the "Environmental Impact" dashboard that has a real-time display of environmental impacts, across the impact categories and life cycle stages, as well as the financial impact. At the bottom of this window, the user has the option to pick the calculation method, choosing between TRACI and CML. Instead of displaying the results for each impact category as total values, they are broken down graphically by percent. A pie chart for each impact category shows what percent each life cycle stage makes up of the total impact. Each impact category, along with the LCC at the bottom, compares percent change, either to the previous design or established baseline set by the user. The easy graphical indication of each impact category updates live with each change and allows the user to quickly understand how each decision affects the environmental profile of an object.

8.5.1.5   Assembly File: Assign Properties

The process for defining the life cycle impacts of an assembly file is similar to that of a part file. Upon first opening the SolidWorks Sustainability task pane, the user is presented with a "Task List" of any part files that are missing material definitions, if there are any. The user is given the option to set the material, and include all of the manufacturing details, or to exclude that part from the overall study. After all parts have a material definition, the user needs to input information about the assembly as a whole, starting with information about the assembly process. The user must input information about the region of assembly and how long the assembly is built to last. Similar to the case of the materials assigned to a whole building assembly, if a part is built to last for less time than the assembly, it will be counted more than once in the LCA. The user can also choose whether or not to include the impacts of the energy required in the assembly process.

The next section asks for information relating to the use of the assembly. This starts with the region where the assembly will be used and any operational energy needs of the assembly over its life span. The use region determines the energy mix used by the product, if applicable, and the end of life conditions. In the next section, "Transportation," default values are assumed based on the region of assembly and use. Lastly, the assumptions about the end-of-life disposal for the assembly, based on percent recycled, incinerated, and landfilled, can be updated if they are known.

8.5.1.6 View Results

The "Environmental Impact" dashboard for an assembly file looks identical to that of a part file. It displays information across the same life cycle stages and impact categories. Users can still switch between the TRACI and CML LCIA methodologies and can still set the duration of use of the product. The same graphs and percent change indicators, as the current design relates to either the previous design or an established baseline, quickly visualize the results of the study.

The "Assembly Visualization" feature of SolidWorks can integrate with the information from SolidWorks Sustainability to visually indicate the impacts of the discrete parts that make up the assembly. This feature can sort the parts and indicate the results by color coding the parts in the order of their impacts. For example, parts of an assembly shown in red contribute the most to GHG emissions and should be focused on to target reductions in total carbon. Some of the properties that can be used with the assembly visualization include carbon footprint, energy consumption, air acidification, water eutrophication, durability, manufacturing location, and manufacturing process. These different ways of viewing the model can help the user to easily identify hotspots in the design and find targeted areas of improvement.

While there is a lot of functionality to view and act on the results of the LCA study within the SolidWorks Sustainability dashboard, there is also the functionality to save the results of the study and further analyze them. At the bottom of the task pane is a button to save the report with three options, as a report (.docx), a spreadsheet (.csv), or a GaBi input file (.xml) which integrates with the GaBi LCA software. The report that is generated gives general information about the product, including the weight, as well as all of the information about manufacturing, use, transport, and end of life that the user inputs while defining the components. The same graphs, one for each impact category broken down by life cycle stage, are also included. Lastly, there is a list of the ten parts that contribute the most to each of the impact categories. The spreadsheet report lists the input assumptions and results by impact category and life cycle stage. Examples of how SolidWorks Sustainability can be used to inform design decisions can be found in Sect. 9.4.

8.6 Comparison of LCIA Tools

The different tools addressed here all answer the same questions and are built on the same scientifically backed framework. That being said, they each offer different functionalities that are better suited to different situations. It is important for designers to know about the different tools in order to find the best option to bring LCA into their own workflows.

8.6.1 WBLCA Tools

There are obvious similarities in the three WBLCA programs driven by the nature of a WBLCA study as well as the standards that guide them. First and foremost, the tools are looking at the same environmental impact categories across the same major life cycle stages. This stems from all of the tools using the TRACI impact assessment methodology for North American studies. Yet there is still variability in the specific stages included (Fig. 8.18). Stages A1–A3 are included in every study and are most important for any high material, long-lasting components like building structure. The use phase impacts, B2–B5, are most important when looking at materials with frequent maintenance and replacement (De Wolf et al. 2017). The other life cycle stages, construction process, end of life, and beyond building life, data is much less reliable as well as more dependent on project-specific conditions and

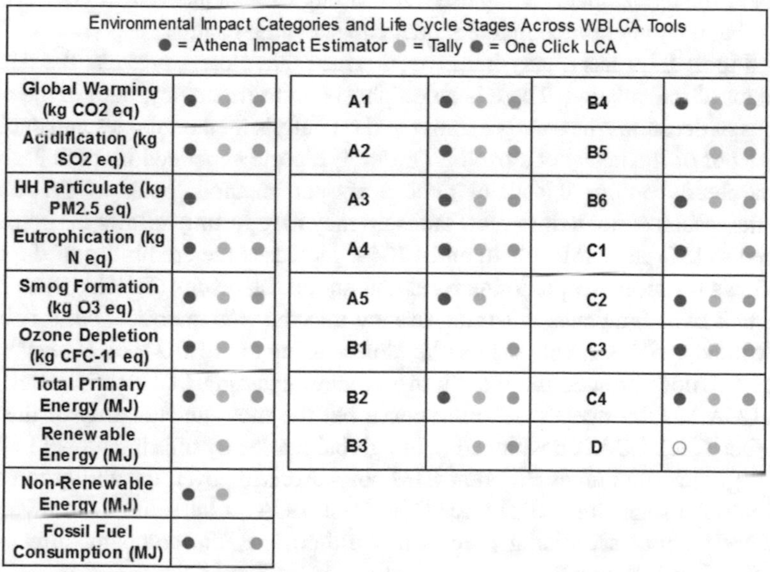

Environmental Impact Categories and Life Cycle Stages Across WBLCA Tools ● = Athena Impact Estimator ● = Tally ● = One Click LCA											
Global Warming (kg CO2 eq)	●	●	●	A1	●	●	●	B4	●	●	●
Acidification (kg SO2 eq)	●	●	●	A2	●	●	●	B5		●	●
HH Particulate (kg PM2.5 eq)	●			A3	●	●	●	B6	●	●	●
Eutrophication (kg N eq)	●	●	●	A4	●	●	●	C1	●		●
Smog Formation (kg O3 eq)	●	●	●	A5	●	●		C2	●	●	●
Ozone Depletion (kg CFC-11 eq)	●	●	●	B1			●	C3	●	●	●
Total Primary Energy (MJ)	●	●	●	B2	●	●	●	C4	●	●	●
Renewable Energy (MJ)		●		B3		●	●	D	○	●	
Non-Renewable Energy (MJ)	●	●									
Fossil Fuel Consumption (MJ)	●		●								

Fig. 8.18 Comparison of WBLCA programs by impact categories and life cycle stages

introduces a lot of variability to studies. Even though all three tools include stages C2–C4, they may have different ways of handling these stages. Furthermore, all three programs use a black-box approach to the majority of the LCA process outside of the LCIA phase. Most of the work for the early stages is done by the programs behind the scenes.

Beyond the ways in which the tools function similarly, there are relative advantages and disadvantages to each of the programs. Athena is the most accessible tool both cost-wise and in terms of compatibility with existing design workflows. Athena is free for all users, and because the user manually enters all of the information about the building constructions and material quantities, it can integrate with any design workflow. It does not matter if the project was completed by hand, using any CAD program, or any BIM program. Athena also requires less specificity to the material entries than the other programs, in favor of rule of thumb sizing, making it easier to use early in the design process. The earlier in the design process that life cycle thinking is applied, the easier it is to make meaningful improvements to the design (Hester et al. 2018). The tradeoff of this is that Athena then offers the least precise results of all three tools. The "Extra Materials" option in each assembly group is a way to combat this. It allows users to add more precision to the study by individually specifying additional materials that are missing from the assemblies.

Since Athena does not interface with any BIM software, as updates are made to the project design, the user must go through and adjust all of the material quantities and constructions manually, making it more time-consuming. Integration with BIM is something that considerably drops the amount of time it takes to complete an LCA study as well as makes assessing design options much easier (Bruce-Hyrkäs et al. 2018). The capabilities and accuracy of these studies are directly tied to the quality of the model that is being used. A thorough, clean model will yield accurate results, but a messy or incomplete model can skew the results.

As a tool, Tally has a very intuitive interface that clearly presents the designer with a lot of information. There is an emphasis on transparency, both in the consequences of decisions the user is making and on Tally's methodologies and decisions that are out of the designer's hands. The PDF reports generated by both Tally and Athena clearly outline the major parameters and methodologies of the study to make the results easier to interpret. Transparency in reporting follows the guidelines outlined in ISO standards 14040 and 14044, increases the credibility of the study, and allows for more people to interpret and act on the results (Frischknecht 2004). One Click LCA integrates this transparency into the web interface, but it is harder to download and report with the results, unless using the LEED tool. However, One Click LCA does provide the most transparency regarding LCI data. Overall, One Click LCA has the steepest learning curve but the most functionality of the three tools. One Click LCA is positioned in the global market by offering global LCI data and integration with many different BIM software. Moreover, the ability to use the web interface standalone allows for One Click LCA to integrate with any design workflow without sacrificing precision. Furthermore, the company offers more methods of evaluating a design than just LCA including LCC.

All of the tools generate the results of the study, presented in numerous ways to help the user interpret them and act on them. However, there are other ways that the tools work to make the results more actionable for the user. The tools dovetail in with several green building standards as they relate to LCA. In the "Reports" window, Athena has a "Rating System Reports" that enables the user to generate reports specific to Green Globes and LEED. One Click LCA has the ability to support results for numerous international certifications, including LEED, Green Globes, and Living Building Challenge. This is important because a 2016–2017 study showed that the overwhelming majority of professionals (mostly in Europe) believe the purpose of a WBLCA study is as part of a green building certification (Bruce-Hyrkäs et al. 2018).

Beyond certifications, the results become actionable when there is some form of comparison that can be made. There is no typical baseline building or target emissions to compare a project against. The existence of such reference standards would assume that all WBLCA studies are completed using the same methodologies and assumptions. The sheer number of variables in methodologies that can be used in conducting a study leads to inherent incompatibility in comparing results across studies. Furthermore, project characteristics including building height, building type, and stories below grade all have considerable effects on the embodied carbon of a project (Simonen et al. 2017). Instead, a more reliable option is to compare the design to variations of itself. Each tool has different methods built in to easily compare different design options to encourage users to compare and evaluate several design options and find solutions with lower impacts on the environment. These comparisons must be done between studies using the same tool and with the same parameters. In most cases, comparisons are only valid when made between options within the defined project scope.

8.6.2 SolidWorks Sustainability

Since SolidWorks is a tool for product designers and engineers, the LCA functionality associated with it requires more technical input regarding materials and manufacturing processes from the user than does a WBLCA software. When looking at the scale of a whole building, the user is removed from any of the manufacturing details of the LCI data they are using. All of this is embedded in the LCI data points that get mapped to building assemblies. In the greater scheme of the study, small differences in manufacturing processes will not have dramatic impacts on the results. However, when looking at the LCA results of a single product, the specifics of the exact material being used in addition to the manufacturing process become much more important to the results of the study. In using SolidWorks Sustainability, users are more engaged with the manufacturing process, fuel sources, fuel quality, and many other technical considerations (Tzetzis and Symeonidou 2015). Furthermore, SolidWorks Sustainability brings LCC right into the LCA dashboard.

The user gets instant feedback on both the environmental and financial implications of every decision that is made. This can help them to optimize both at once, so that one evaluative criteria do not suffer at the expense of the other.

8.7   Other LCA Tools

While this chapter covered four common LCIA tools in North America aimed at designers and other non-LCA experts, there are dozens of other tools that can be used all with their unique objectives, strengths, and weaknesses. Different tools will suit different practitioners based on many factors including their level of familiarity with LCA, the objective of their study, underlying familiarity with different computer programs, and financial resources. This list is nowhere close to exhaustive but aims to show the range of options available when selecting an LCA tool.

8.7.1   Rhino.Inside.Revit

LCA studies rely on a lot of different inputs of a variety of information, including materials, building assemblies, and quantities. Due to the nature of a BIM model and the amount of information embedded within them, they pair well with LCIA tools. However, other 3D modeling software, such as Rhinoceros 3D (Rhino), only have information about geometry, making it very hard to interface with an LCA tool. While there are currently no LCA tools that can interface directly with Rhino, a new beta from Robert McNeel & Associates and Autodesk can help to bridge this gap and bring the power of LCA to more design workflows. Rhino.Inside.Revit was not designed specifically for LCA studies but instead to create a way to link Revit and Rhino models, using Grasshopper, to combine the power and workflows of the tools. This enables users to bring geometry from Rhino into Revit as native Revit elements instead of just a generic massing like in the past. In doing so, the elements can then interface with all of the same LCA plugins that a model built originally in Revit can. Anything built in Rhino, which originated as just geometry, can now contain all of the embedded information that is paramount of a BIM model.

Rhino.Inside.Revit comes with specialized Grasshopper components that can be used to rebuild the model in Revit rather than just transfer the geometry. These "Revit Build" components require the same inputs and parameters that are required when building a model in Revit so that any geometry transferred becomes an element native to Revit. For example, if a floor slab is brought from Rhino to Revit, the *Add Floor* component will be used so that Revit knows what type of building component the element is. Furthermore, the user adds all of the parameters related to the floor construction, materials, and associated level. Using all of this information, a full floor component is built in Revit but is linked to the Rhino model and will update with any changes made there. With a very simple grasshopper script and a little bit of time, a whole Rhino model can be translated to Revit and then interfaced with any of the available LCA plugins. Furthermore, just like Dynamo, it gives the user the ability to set up fully parametric Revit models, which makes it quick and easy to test different design and material configurations, and then get LCA results for these steps. Overall, Rhino.Inside.Revit opens up a whole host of new opportunities to easily integrate Revit-specific LCA studies into Rhino workflows, which was not previously possible. For more information, please visit https://www.rhino3d.com/inside/revit/beta/.

8.7.2   NIST BEES and BIRDS

Among the software released by NIST, BEES and BIRDS stand out as sustainability solutions for architects and designers. Both of these tools aim to take a more holistic approach to sustainability and incorporate elements of life cycle costing, indoor environment quality, and operational energy in addition to LCA (Kneifel and O'Rear 2018). The two tools differ in terms of their object of study and evaluative criteria but are both provided as free web-based tools. The BEES framework addresses environmental and economic impacts of individual building products, while BIRDS provides a framework to evaluate whole buildings across environmental impacts, economic impacts, indoor environment quality (IAQ), and operational energy.

BEES is a tool that can help designers compare individual building products by environmental and economic impacts. The interface is simple with a very small learning curve and is designed to be able to be used without a lot of background knowledge on LCA (Kneifel and O'Rear 2018). Building components with available information are organized by building component groups and type according to the UNIFORMAT II standard (Kneifel et al. 2018). The user has the ability to select various criteria to evaluate the product on, including commercial versus residential applications, the LCIA method used, how much material is being used, and the discount rate for the LCC component. From an available list of building products, the user picks one to set as the benchmark and other products to evaluate against the chosen benchmark. The advantages of an LCA study, as discussed, are the ability to evaluate across impact categories to look for true reductions instead of burden shifting. This can be challenging to do, though, as the different impact categories are not comparable. For more information, please visit https://www.nist.gov/services-resources/software/bees.

BIRDS is a whole building sustainability framework also developed by NIST. While BEES focused on just two evaluative criteria at the scale of an individual product, BIRDS aims to tackle all of the sustainable criteria defined by NIST and the tradeoffs between them. There is a strong link between operational energy, embodied environmental impacts, and total cost that BIRDS attempts to tap into (Kneifel and O'Rear 2018). In order to do this, BIRDS is limited in the ability to customize the inputs. After selecting either a commercial or single-family detached residential building, BIRDS then provides choices to the users regarding building type (only for commercial) and relative size. These are used as stand-ins for the building and determine the building size, material constructions, and HVAC equipment. The user then has the ability to select different building locations, energy codes, and length of study. The user can test multiple options for each of these variables to compare results. While the prototype buildings included in BIRDS represent the typical building typologies, many factors including material selection, geometry, and operational conditions can have a material impact on the results of these studies, and therefore the results from BIRDS may not be representative of the specific building that is being evaluated (Kneifel and O'Rear 2018). For more information, please visit https://www.nist.gov/services-resources/software/birds.

Both BEES and BIRDS utilize an environmental impact score (EIS) in the results of the studies that creates a combined weighted total of the impacts across the chosen impact categories included in the study (Kneifel and O'Rear 2018). The user can pick from predefined weights, such as those from the US EPA or BEES, or define their own weighted average. The idea of including a single score is to make the comparison between two or more options spanning several impact categories easier. However, it is also something that adds a lot of bias to the study and is prohibited in ISO standards 14040 and 14044. LCA is meant to be utilized as an evaluative tool and does not claim to provide a definitive answer. An EIS aims to bring an answer to a study, but based on what weighting is chosen, this answer will be different for different practitioners. On the other hand, it makes it so that the user of the tool, who will often not be a chemist or environmentalist, is not trying to weigh the impacts in different categories at random. Instead, the predefined weighting choices are built on science and expert opinions and provide a more sound, albeit biased, point to make a decision from.

8.7.3   SimaPro and Other LCA Tools

All of the LCIA tools included up until this point have been aimed at designers and other non-LCA experts. However, there are other tools that allow users to be involved with every step of the LCA process. These tools, aimed at people with lots of knowledge about LCA, have a much steeper learning curve but allow for a much more detailed study. One such example is SimaPro, an LCA tool developed by PRé Sustainability. Another example, GaBi is run by Sphera. Partially due to their international applicability, these are the two of the most commonly used LCA software (Speck et al. 2015). Since they are aimed at LCA professionals, both in industry and academia, they offer full control and transparency to the user in terms of methodological decisions and underlying assumptions at all steps of the LCA process. While a majority of the design tools discussed at length in this chapter are black-box types with the underlying data out of the view and control of the designer, LCA tools bring the user into every stage and decision of the study. This level of control allows the user to fully tailor the study to meet their needs but also requires more time and knowledge of underlying LCA principles.

When using SimaPro, the interface is divided into four sections for the phases of an LCA study. The user can then go through all of the stages of a typical study. Under goal and scope, the user has space to document the study as well as select what databases they want to pull from. As discussed in Chap. 7, several LCI databases feed into SimaPro. The LCI tab is where the user can add any processes and life cycle stages to the assemblies. The databases come with preloaded processes that show all of the inputs and outputs of one set process within the study. However, SimaPro also allows the user to create their own. These processes can then be applied to assemblies and life cycle stages to complete all of the flows that make up the object of study. After all of this information is added, the user then has control over what LCIA method is used. Filling out all of these required parameters brings the user to graphical and tabular results of the study. Results can be viewed across impact assessment categories, as a single impact score like in BEES and BIRDS. Overall, this level of control and detail requires more information from the user but results in a more accurate study. For more information, please visit https:// simapro.com/.

GaBi operates using three main components, a plan, processes, and flows. Creating a new plan establishes the life cycle of the product or system being studied and hosts the model. Then, processes are added to represent the different stages of production. These can be pulled from the extensive list of predefined processes, or the user can make their own. Flows are used to represent the movement of matter and energy, and they can connect processes as well as represent inputs and emissions. GaBi has a lot of built-in functionality to compare multiple scenarios and best improve the design. Results can be viewed right on the model, just like the ability to see Athena's results on the project tree, or in more detailed graphical format. Results can be viewed for the overall model as well as an established group of processes reflecting a portion of the overall model.

8.8   Conclusions

The American Institute of Architects (AIA) Framework for Design Excellence compares WBLCA to an energy model for a building's materials (AIA n.d.). This lends itself to borrowing a common phrase in energy modeling "garbage in, garbage out." The same is true of LCA studies. If designers do not understand what they are doing when using these tools, then a study can inadvertently influence the designer to make worse choices for the environment, which is opposite the goal. Yet, with the proper understanding of these tools and the LCA processes that undergird them, they are powerful evaluative methods in the hands of designers to make decisions that will reduce the net negative environmental impacts across multiple categories.

References

AIA (n.d.) Framework for design excellence. https://www.aia.org/resources/6077668-framework-for-design-excellence. Accessed May 2020

Aktas CB, Bilec MM (2012) Impact of lifetime on US residential building LCA results. Int J Life Cycle Assess 17:337–349. https://doi.org/10.1007/s11367-011-0363-x

Athena Sustainable Materials Institute (n.d.) www.athenasmi.org. Accessed June 2020

Bare J (2012) Tool for the reduction and assessment of chemical and other environmental impacts (TRACI) user's manual. United States Environmental Protection Agency. https://nepis.epa.gov/Adobe/PDF/P100HN53.pdf. Accessed July 2020

Bionova (n.d.) One Click LCA. www.oneclicklca.com. Accessed June 2020

Bowick M, O'Connor J, Meil J (2017) Whole-building LCA benchmarks. A methodology white paper. Athena Sustainable Materials Institute. http://www.athenasmi.org/wp-content/uploads/2017/11/BuildingBenchmarkReport.pdf. Accessed May 2020

Bruce-Hyrkäs T, Pasanen P, Castro R (2018) Overview of whole building life-cycle assessment for green building certification and ecodesign through industry surveys and interviews. Procedia CIRP 69:178–183. https://doi.org/10.1016/j.procir.2017.11.127

Bulle C, Margni M, Patouillard L et al (2019) IMPACT World+: a globally recognized life cycle impact assessment method. Int J Life Cycle Assess 24:1653–1674. https://doi.org/10.1007/s11367-019-01583-0

Carbon Leadership Forum (2020) Wood carbon seminars, session 2: LCA and wood, 2.1 – carbon neutrality [webinar]. https://carbonleadershipforum.org/projects/wood-carbon-seminars/

Cays J (2017) Life-cycle assessment: reducing environmental impact risk with workflow data you can trust. Archit Des 87:96–103. https://doi.org/10.1002/ad.2179

CML (2016) CML-IA characterisation factors. https://www.universiteitleiden.nl/en/research/research-output/science/cml-ia-characterisation-factors. Accessed July 2020

Curran MA (ed) (2012) Life cycle assessment handbook: a guide for environmentally sustainable products. Scrivener Publishing/Wiley, Salem/Hoboken, pp 23, 78–80, 83–86

Dassault Systèmes (n.d.) SolidWorks Sustainability. https://www.solidworks.com/sustainability/. Accessed June 2020

De Schryver AM (2011) Value choices in life cycle impact assessment. Dissertation, Radboud University

De Wolf C, Pomponi F, Moncaster A (2017) Measuring embodied carbon dioxide equivalent of buildings: a review and critique of current industry practice. Energ Buildings 140:68–80. https://doi.org/10.1016/j.enbuild.2017.01.075

Dekker E, Zijp MC, van de Kamp ME et al (2019) A taste of the new ReCiPe for life cycle assessment: consequences of the updated impact assessment method on food products LCAs. Int J Life Cycle Assess. https://doi.org/10.1007/s11367-019-01653-3

Frischknecht R (2004) Transparency in LCA- a heretical request? Int J Life Cycle Assess 9:211–213. https://doi.org/10.1007/BF02978595

Goedkoop MJ, Spriensma R (2000) The Eco-indicator 99 a damage oriented method for life cycle impact assessment methodology report. PRé Consultants BV. http://www.pre-sustainability.com/legacy/download/EI99_annexe_v3.pdf. Accessed July 2020

Hauschild MZ, Goedkoop M, Guinée J et al (2013) Identifying best existing practice for characterization modeling in life cycle impact assessment. Int J Life Cycle Assess 18:683–697. https://doi.org/10.1007/s11367-012-0489-5

Hester J, Miller TR, Gregory J et al (2018) Actionable insights with less data: guiding early building design decisions with streamlined probabilistic life cycle assessment. Int J Life Cycle Assess 23:1903–1915. https://doi.org/10.1007/s11367-017-1431-7

Huijbregts MAJ, Steinmann ZJN, Elshout PMF et al (2016) ReCipe 2016 v1.1: a harmonized life cycle impact assessment method at midpoint and endpoint level. National Institute for Public Health and the Environment. https://www.pre-sustainability.com/legacy/download/Report_ReCiPe_2017.pdf. Accessed July 2020

Jolliet O, Margni M, Charles R et al (2003) IMPACT 2002+: a new life cycle impact assessment methodology. Int J Life Cycle Assess 8:324–330. https://doi.org/10.1007/BF02978505

King B (ed) (2017) The new carbon architecture: building to cool the climate. New Society Publisher, Gabriola Island, p 33

Kneifel J, O'Rear E (2018) Challenges and opportunities in quantifying and evaluating building sustainability. Technol Archit Des 2:160–169. https://doi.org/10.1080/24751448.2018.1497363

Kneifel J, O'Rear E, Webb D et al (2018) Building industry reporting and design for sustainability (BIRDS) building code-based residential database technical manual: update. National Institute of Standards and Technology. https://nvlpubs.nist.gov/nistpubs/TechnicalNotes/NIST.TN.1999.pdf. Accessed July 2020

KT Innovations (n.d.) Tally. www.choosetally.com. Accessed June 2020

Monteiro H, Freire F (2012) Life-cycle assessment of a house with alternative exterior walls: comparison of three impact assessment methods. Energ Buildings 47:572–583. https://doi.org/10.1016/j.enbuild.2011.12.032

Rodriguez BX, Simonen K, Huang M et al (2019) A taxonomy for whole building life cycle assessment (WBLCA). Smart Sustain Built Environ 8:190–205. https://doi.org/10.1108/SASBE-06-2018-0034

Simonen K, Rodriguez BX, De Wolf C (2017) Benchmarking the embodied carbon of buildings. Technol Archit Des 1:208–218. https://doi.org/10.1080/24751448.2017.1354623

Speck R, Selke S, Auras R et al (2015) Life cycle assessment software: selection can impact results. J Ind Ecol 20:18–28. https://doi-org.libdb.njit.edu:8443/10.1111/jiec.12245

Tzetzis D, Symeonidou I (2015) Material and design selection of wine packaging using a CAD-based approach for green logistics. Presented at the 1st international conference on agrifood supply chain management & green logistics, Porto Carras Grand Resort Halkidiki, Greece, 27–30 May 2015

Yazdanbakhsh A, Bank LC, Baez T et al (2018) Comparative LCA of concrete with natural and recycled coarse aggregate in the New York City area. Int J Life Cycle Assess 23:1163–1173. https://doi.org/10.1007/s11367-017-1360-5

Chapter 9
Case Studies

Abstract Using analytical tools is a good first step to making better data-driven design decisions. The second, even more important, step in the process is interpreting and acting on the results. The results of the LCIA phase of an LCA study will rarely produce a clear "winning" design option that is obviously better than the others under consideration. A properly framed and conducted study does, however, provide a clearer picture of the environmental impacts associated with various design component. It is then up to the designer to weigh the pros and cons of each decision, guided by a clearly stated goal and scope of the study. This chapter will address both the ways in which each of the LCIA tools best function as well as how the results of the tools can be used to affect positive change in design decisions at the building component scale. The general workflow followed in the case studies here can be applied to many other design disciplines working at smaller and larger scales.

All of the LCIA tools are founded on the same scientific framework, introduced in Chap. 5; pull from overlapping data sources, presented in Chap. 7; and use the same LCIA method, described in Chap. 8. Even with the differences in methodologies used by each tool, which make studies completed in different programs *not* comparable, there is overall common agreement in the results of studies completed across multiple tools looking at the same set of design elements. In most cases each tool will lead a user to the same general conclusions regardless of what tool is being used.

9.1 The MIA

The module for impact assessment (MIA) is a small, theoretical project that serves as a framework of different building constructions and material choices. The MIA is a two-story box, roughly 25′ × 40′, with only a structural system and exterior envelope. The structure is modeled after Le Corbusier's *Maison Dom-Ino* (Foundation Le Corbusier n.d.) (Fig. 9.1). The MIA was used to analyze the results of the three WBLCA tools and determine what similarities and trends exist across the results of the different tools. The project is assumed to be located in the Northeast United States, and a 100-year service life was used for the project. The results are all calculated for life cycle stages A–C and omit module D. One Click LCA does not

© Springer Nature Switzerland AG 2021   175
J. Cays, *An Environmental Life Cycle Approach to Design*,
https://doi.org/10.1007/978-3-030-63802-3_9

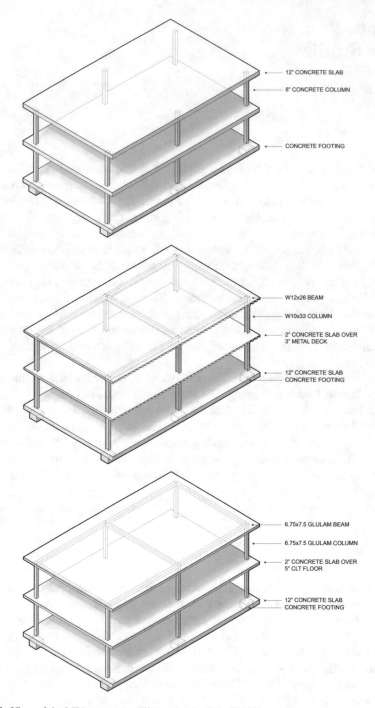

Fig. 9.1 View of the MIA structure. (Illustration by Erin Heidelberger)

provide results for this stage, Athena has the option to include them, and Tally automatically adds them in. However, they are not frequently included in LCA studies due to the lack of consensus on how to calculate and integrate the results to the study (De Wolf et al. 2017).

9.1.1 Functionality of LCIA Tools

In order to best understand the trends in the results, LCA studies were carried out for discrete components of the MIA and compared at the level of a single building construction rather than a whole building. The structure, roof construction, exterior wall, and interior wall finishes were all analyzed separately, with different material choices, to compare results. The three structural systems studied were a concrete frame, a steel frame, and a wood frame. The concrete structure includes six concrete footings, a concrete slab on grade, six 8″ square concrete columns, and two 12″ concrete floor slabs. The steel structure includes six concrete footings, a concrete slab on grade, six W10×33 steel columns, 14 W12×26 steel beams, and two floor slabs made up of a 2″ concrete slab over 3″ metal decking. The wood structure includes six concrete footings, a concrete slab on grade, six 6.75″ × 7.5″ glulam columns, 14 6.75″ × 7.5″ glulam beams, and two floor slabs made up of a 2″ concrete topping over 5″ CLT plank.

Seven different exterior wall constructions were studied, all with differing R-values. The FU for the wall study was 147′ × 22′, or 3234 sf, of wall. The wall constructions tested are a CMU wall, a precast concrete wall, a masonry wall, brick cladding over a metal stud wall, steel panels over a metal stud wall, brick cladding on a CMU wall, and wood siding on a wood stud wall. Five different interior wall finishes were studied. They were painted gypsum, painted gypsum with furring channels, wall tiles, wall tiles over furring channels, and plaster over metal lath. This study looked at the trends between different material choices for one building construction using Athena, Tally, and One Click LCA.

Based on the nature of the way that Athena collects information about the project materials being used, it is not possible to fully isolate these studies. Athena uses the major assembly groups to input information about a project and hosts secondary information within them. To get information on just the roof construction and wall finishes from Athena, a model with the host component, either the roof slab or wall, was used. Results were calculated with and without the added "envelope" construction in each case, and the difference in the results reveals the impacts of just the envelope component. Tally and One Click LCA have more flexibility to isolate individual components and run an LCA study on a discrete part. Anything that can be modeled in Revit can be the object of study in Tally and One Click LCA. The roof construction and wall finishes were modeled as independent families, and the study was conducted on the isolated construction.

9.1.2 Results

Looking first at the results of the three structural systems, GWP shows the same trends across all three tools (Fig. 9.2). Each tool shows concrete having the highest impact and wood having the lowest. All three tools point to wood being the clear choice for the structural system when looking only at GWP, but depending on what tool is used, the decision between concrete and steel gets more difficult to make based solely on one impact category. However, the interpretation of the results of a study is never completed in the vacuum of one impact category but rather requires the user to balance the results across many of them.

While the results of the tools are largely in agreement, there are instances where the selection of a tool will impact the results and lead to different conclusions (Herrmann and Moltesen 2015). The reasons for these variations in the tools can sometimes be attributed to differences in regional versus location-specific data used by the tools or transportation assumptions that tool assumes (Sinha et al. 2016). Most commonly, these differences are caused by varying or missing CFs, even when using the same LCIA methodology (Speck et al. 2015). It is important to note that the differences are not caused by the LCI data but rather the overall implementation of the LCIA methods and general methodologies of the tools themselves. In full LCA tools, the differences in results can be traced back through the process and pinpointed (Herrmann and Moltesen 2015). With the lack of transparency in LCIA design tools, the user cannot determine where the differences lie.

Looking beyond the GWP of the MIA, there is less agreement across the three tools in the case of acidification (Fig. 9.3). Each of the tools shows a different frame

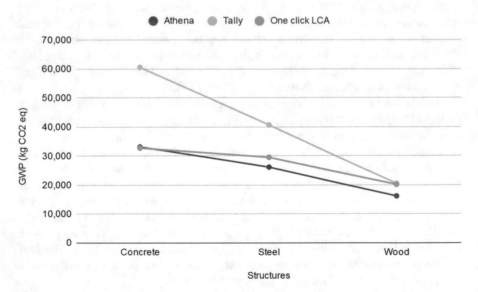

Fig. 9.2 Trends across WBLCA tools for three structural systems. Results are for global warming potential (in kg CO_2eq) for life cycle stages A–C

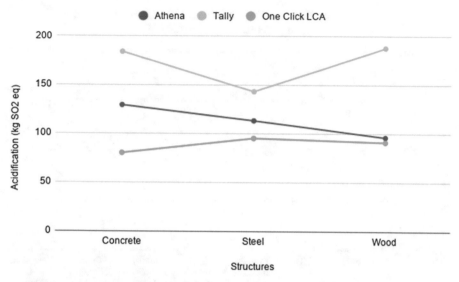

Fig. 9.3 Trends across WBLCA tools for three structural systems. Results are for acidification (in kg SO₂eq) for life cycle stages A–C

material as having the lowest acidification potential, and each tool has a varying amount of difference between the tools. For instance, between steel and wood, One Click LCA only shows a 4% decrease, Athena shows a 15% decrease, but Tally shows a 31% increase. Each of these tools shows a different relationship between the acidification of the structural materials. However, the total numbers across the three materials in each of the three systems are close together. It can be reasonably stated that these results, combined with the other impact categories, would not lead to a different conclusion when using different tools.

Shifting to look at the trends in the results of the exterior wall constructions, there is again agreement in the trends of GWP (Fig. 9.4). Athena has LCI data for different types of brick cladding, but no data points are available on a masonry wall construction, so there is no value included there. A notable area that the graphs do not track as closely is surrounding the CMU wall construction. After investigating the material assumptions for each of the three programs for the CMU blocks, it was determined that Athena and One Click LCA had an assumed density of 2002 kg/m³, or 125 lb/ft³, for the CMU block. However, Tally had a default assumed density of 927 kg/m³, or 58 lb/ft³, contributing to the differences. The density of the block can be manually updated in Tally to match the other programs, but the point of this exercise is not to get the results to match but rather to see what decisions each of the three tools would lead someone to make when used independently. This points to the importance of making sure the LCI data selected matches the materials used in the project.

One notable area where the tools would lead to different conclusions is specifying steel panel walls in One Click LCA (Fig. 9.5). As can be seen in the first graph,

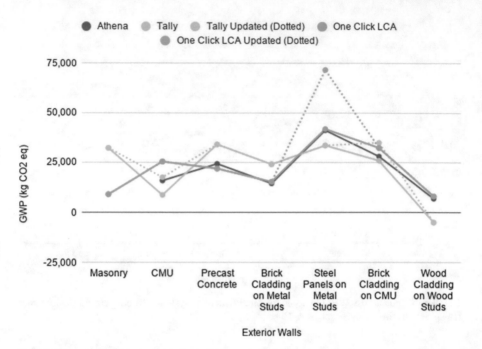

Fig. 9.4 Trends across WBLCA tools for seven exterior wall constructions. Results are for global warming potential (in kg CO_2eq) for life cycle stages A–C

the initial value for acidification was an order of magnitude larger than any values for the other wall assemblies or tools. This initial data point was manufacturer specific, but no specific product for this project was chosen. Changing the LCI dataset used from this to a generic dataset, from NREL, brings the acidification levels much closer to the anticipated level based on the other tools and materials studied. However, this also pushes up the GWP of the steel panel even higher. Overall, the selection of the tool can lead to a different decision in the case of the steel wall panel. Typically, the tools lead to the same conclusions, but there are some exceptions (Herrmann and Moltesen 2015). Overall, increased standardization will contribute to more agreement across the tools, but in the majority of cases, it already exists.

Interior wall finishes were also looked at in isolation in this study to show the different consideration that goes into finishes as opposed to building structure and envelope. Interior finishes account for a much smaller percentage of the initial building mass and impacts than the envelope and structure do. However, finishes also have a much shorter service life than the other components and will be replaced frequently throughout the life of the project. This drives the use stage life cycle impacts up for the finish materials. The product service life is both more important and harder to estimate for finishes because they are often replaced based on changing styles not because of functional deficiencies (Aktas and Bilec 2012). Finishes, while sometimes included in studies, are not one of the major categories typically

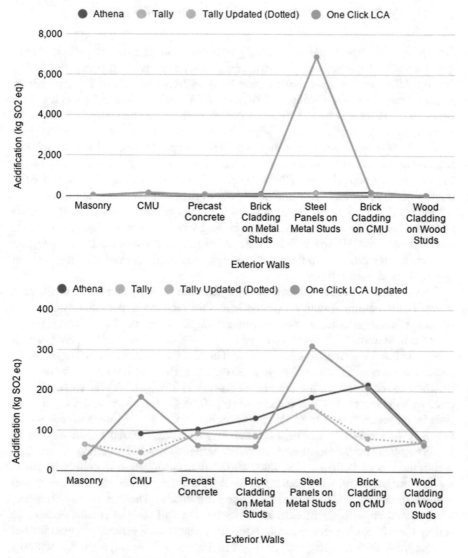

Fig. 9.5 Trends across WBLCA tools for seven different exterior wall constructions. Results are for acidification (in kg SO₂eq) for life cycle stages A–C

included in WBLCA studies. This is reflected by the fact that Athena does not have any data to look at a tile wall, and One Click LCA did not have any North American data to look at a plaster wall. Across the three programs, LCI data on finish materials is far more limited than more common WBLCA elements. The full table of results for two configurations of multiple higher mass building construction assembly types, impact category, and life cycle stage can be found in Appendices A and B.

9.1.3   Environmental Versus Aesthetic Considerations

Environmental impacts are not the only consideration in a design process. Many quantitative and qualitative considerations go into any design process. Just like it can be difficult to balance benefits and burdens between different, incompatible impact categories when using one of the LCIA tools, balancing considerations between different assessment methods proves even more challenging. There is no right or wrong answer, only an informed conclusion based on all of the evidence.

Taking the same framework setup in the MIA, a qualitative consideration of ceiling height can be added to the conversation. The first default configuration of the MIA had an 8′ ceiling height. What happens if this is increased to 12′ to improve the quality of the space and the overall aesthetics of the design or to accommodate typical mechanical systems? If the proposal was being evaluated purely based on environmental impacts, or even cost, the 8′ ceiling would be the obvious choice as it uses less material. However, the experience of being in a space with 12′ or higher ceilings is very different to that of 8′ ceilings, and architects and designers cannot ignore these considerations.

The increase in ceiling height affects the different building categories in different ways. The structural system only experiences an increase in column sizing due to the height change, whereas the exterior envelope experiences a more significant increase in materials. The total wall area of the MIA increased by about 36%, which is reflected in the change in total impacts. The GWP and acidification for the walls increased anywhere from 16% to 40% across all of the material configurations and the three tools. However, the average increase was 35%, which very closely correlates with the increase in area of the wall (Fig. 9.6). On the other hand, the structure only increased anywhere from 1% to 5%, for the concrete structure in Athena which saw a 10% increase. This change is much less significant than the change found in the envelope. The increased environmental impacts of a taller ceiling do not outweigh the design benefits. This study shows that the increase in ceiling height is much more impactful when looking at the envelope than structure. This means that while no changes need to be made regarding the structural choice, careful consideration should be given to the exterior envelope. The full table of results, updated to reflect the 12′ ceiling, for multiple building construction assembly types, impact category, and life cycle stage can be found in Appendix B. Average impact increases in GWP and acidification of the MIA exterior wall and structural types going from an 8′ to a 12′ floor to ceiling can be found in Appendix C.

9.2   Building 197

Originally built as a last-mile shipping warehouse in Kearny, NJ, Building 197 is an adaptive reuse project to provide an existing office complex additional space to work and gather (Fig. 9.7). The original structure is made up of tilt-up concrete

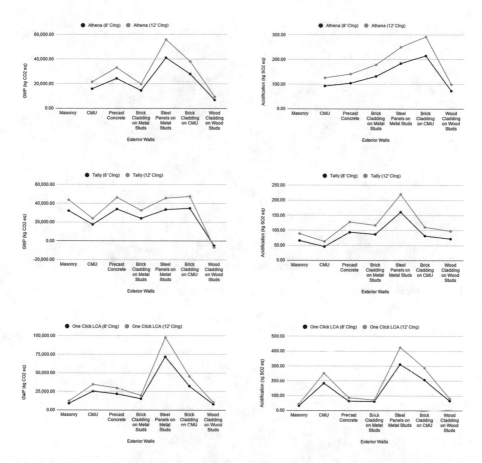

Fig. 9.6 Changes in GWP and Acidification of the MIA exterior walls between an 8′ and 12′ ceiling

panels, steel columns, and open-web joists. The interior space has 36′ clear ceilings and no partitions. The building measures approximately 250′ × 750′.

9.2.1 WBLCA Study

Before looking for any specific components or targeted improvements of the project, it is necessary to establish a baseline WBLCA study to analyze the results of and compare improvements against. Either a baseline design or established benchmark to compare against makes the results of a WBLCA study the most useful (Athena 2020). This project was designed, and materials were selected, before life cycle thinking was applied. This means that no comparisons between design options were made throughout the design process. Instead, this is a retrospective look back at the design decisions made in an attempt to question and quantify assumptions. The project was designed using Revit, and therefore Tally and One Click LCA were

Fig. 9.7   Interior view of Building 197 Erin Heidelberger student project

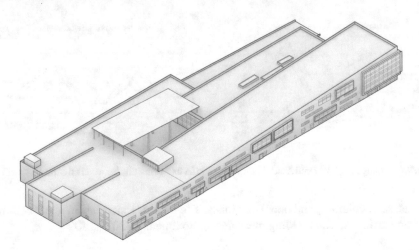

Fig. 9.8   Schematic exterior view of Building 197 Erin Heidelberger student project

used to complete assessments on the project over a 100-year service life looking at cradle-to-grave impacts of the design. The WBLCA study was inclusive of structure, foundation, and envelope (Fig. 9.8). This comprises the vast majority of the building mass. Within these categories, the structure is the dominant element in terms of both mass and embodied impacts. The results of the WBLCA were used to inform all of the ensuing studies and reductions in overall impacts.

As discussed in Sects. 8.3.3 and 8.3.4, a BIM model contains data about a building's construction and performance but does not simulate actual built conditions. Furthermore, a BIM model is only as knowledgeable and useful as the data that is

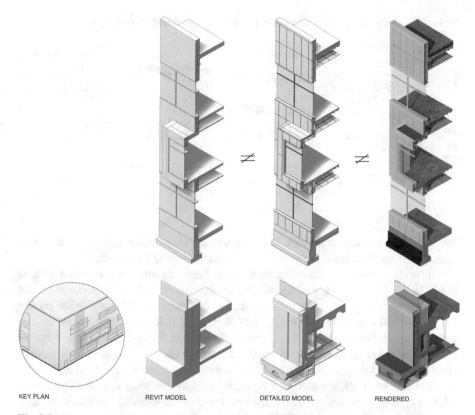

Fig. 9.9 Abstracted BIM model versus more detailed (but still abstracted) models simulating real-world built conditions of Heidelberger student project

added to it. Regardless of how detailed a BIM model is, it will never be the same as what actually gets built (Fig. 9.9). For instance, in most Revit window families, the window frame is represented as a straight rectangular extrusion of the selected material. Similarly, when framing elements, such as light gauge metal studs, are used in a wall construction, Revit models a solid rectangular prism for that layer of the wall.

This is something that each WBLCA tool recognizes and approaches differently. In the case of Tally, each building component has individualized methods of takeoff and additional materials that can be added to bring the BIM model closer to real-life conditions. For example, when assigning information about metal studs to a wall construction, Tally has the user specify the size and spacing of studs for takeoff. Using this option, Tally takes the rectangular prism provided by Revit and fills in that area with metal studs based on the spacing. For other building elements, takeoff is more straightforward, but there are additional materials that would be used in the construction excluded from the Revit model. Every time a concrete element is defined in Tally, the user has the ability to add reinforcing members, such as rebar.

Table 9.1 Comparison of LCA results for Building 197 with and without additional materials in Tally

	Tally-whole Building LCA	WBLCA without additional materials	Percent decrease in impact
Global warming (kg CO$_2$eq)	15,709,168	11,640,089	26%
Acidification (kg SO$_2$eq)	81,719	56,822	30%
Eutrophication (kg Neq)	3164	2585	18%
Smog formation (kg O$_3$eq)	917,372	688,883	25%
Ozone depletion (kg CFC-11eq)	0.05	0.04	20%
Non-renewable energy (MJ)	242,532,107	197,241,102	19%

In the case of a curtain wall door, the user is able to input information not included in the Revit model related to the finish, hardware, hinges, and closer.

While all of these additional materials may seem inconsequential in the greater WBLCA, the resultant sum of them adds up to make a material difference on the results of the study. To show the impact these additional materials make on the study, the WBLCA of Building 197 was completed twice in Tally, both with and without the additional materials. Elements like rebar, spray fireproofing, window and door hardware, or fasteners were the difference between the studies. The only exception to this was in cases where the user was required to select an additional material. For example, when assigning metal wall panels, Tally requires the user to select a type of fastener, with no option to not include them in the study because they are required in the built assembly.

Comparing the results of the two WBLCA shows between 18% and 30% decrease in all impact categories without the additional materials (Table 9.1). These results clearly demonstrate why it is important to understand what data is and is not included in a BIM model as well as what the goal and scope of an assessment is. Every decision made in completing a WBLCA study has the ability to appreciably alter the results. This is why directly comparing two LCA studies carried out separated is not recommended. It is essential to maintain methodological consistency between design option studies carried out by one person or team. Using the same tools and methods for one project will maintain this internal project consistency.

The other WBLCA tools have their own ways of handling the gap between project data and real-world conditions. In the One Click LCA Revit plugin there is an option for the user to add rebar to any concrete elements. Beyond this, information on how the data point was collected and what it reflects can be seen in the EPD used for the LCI data, if one was used. Athena does not interface with any outside programs and so is not drawing from any models. In every building assembly that the user adds, Athena has additional information and materials related to building assemblies that it automatically adds to the study. These additional materials, and their quantities, can be seen in the bill of materials that Athena generates. For example, after adding a wall assembly composed of brick cladding over metal studs, the bill of materials reflects joint compound, mortar, nails, screws, and paper tape, all of which were not specified by the user. In all programs, these additional materials are estimates of real-world conditions that more closely reflect how the building will be constructed, but a WBLCA study will never be inclusive of every single element used to actually build the building.

9.2.2 Structural System Comparison

The results of the initial WBLCA study, inclusive of fasteners and additional materials, were analyzed to find targeted areas of improvement within the project to reduce impacts, specifically paying attention to GWP as this is the impact category highlighted by the tools. The "Most contributing materials" section on One Click LCA revealed that of cradle-to-gate (life cycle stages A1–A3) GWP, the two most contributing materials are the "Steel roof and floor deck" and the "Steel, structural, beams and columns." They account for 36.1% and 24.1% of the impacts respectively. Looking at the results of the WBLCA study completed in Tally broken down by material divisions, the metals division is the largest contributor to GWP over the building's life cycle at 33% of the total. When looking at the totals broken down further by both division and Tally entry, the "Steel, W section (wide flange shape)" alone accounts for 23% of the building's GWP across the life span of the project.

Using these breakdowns of the WBLCA study, the building's structural system is a clear area to target improvements within the project design. A steel structural system was originally selected for the project to tie into the existing steel columns that were salvaged. However, a second steel column was coupled with every existing one to account for the additional loads of the two floors that were added to the project. Furthermore, steel beams and metal decking were added to support the new floors. In the end, the majority of the structural system was new and was a big contributor across the impact categories. In this structural system comparison, the steel system used in the original design was compared against a concrete structural system and a wooden structural system.

This study, over a 100-year service life, was inclusive of only structure, not foundations, envelope, or finishes. The original structural bay sizing, which is larger than usual because it was designed for flexibility in a warehouse, was maintained in the initial design as well as all structural systems tested. The steel system consists of the salvaged steel columns matched with a new, secondary column and steel beams. Spray fireproofing was added to all of the beams, and each column was encased by two layers of gypsum board to meet required fireproofing. The floor system consists of reinforced, cast-in-place concrete over metal deck. The concrete structural system consists of precast concrete columns and beams. The floor system consists of precast hollow core concrete panels topped with a thin, reinforced layer of cast-in-place concrete. The wood structural system consists of glulam columns and beams. The floor system consists of CLT planks topped with a thin, reinforced layer of cast-in-place concrete.

The benefit of an LCA study is the holistic look at environmental impacts across a building's life span in several different categories. This approach ensures that no one category will suffer greatly at the expense of another (Athena 2019). However, this can also be somewhat of a drawback when it comes to acting on the results of the study. While one impact category may indicate that one option is the better choice, a separate impact category may show the opposite. Factors including who is carrying out a study and why may influence which impact categories priority is given to. If this study was only concerned with GWP, then a wood structural system would be the clear choice to use as it has the lowest impact using both Tally and One Click LCA (Fig. 9.10). However, when taking a broader look at the project across impact categories, it gets more challenging to figure out what the best decision to make is.

When looking at the results from One Click LCA, a wood structural system is the least impactful choice in all categories barring eutrophication where it was very close to the values of concrete. In Tally, wood is only the least impactful choice in GWP but is close to the other values in all impact categories except for eutrophication, which stands out significantly on the graph. It is worth noting that the numerical values do not even differ by one order of magnitude. Tally shows 1402 kg Neq and 1247 kg Neq for steel and concrete, respectively, and 8277 kg O_3 for wood. While this difference seems extreme on the graph, looking at the values themselves tells a different story. Eutrophication does set wood apart from the other materials in Tally, but not enough to outweigh the benefits in the other categories.

It can be very hard to negotiate the tradeoffs between environmental impact categories and reach a final conclusion. Several practitioners and tools, including BEES and BIRDS, have recognized the challenge in negotiating the incompatible impact categories, and many attempts have been made to combine the results of all of the different impact categories into one overall score indicative of the total environmental impacts of the design (Kneifel and O'Rear 2018). Further discussion on the benefits and drawbacks of this can be found in Sect. 8.6.2.

Overall, this study tested the assumption made that a steel structural system would be the best choice because it could reuse the existing columns. However, the majority of the steel was new, and this option ended up being more impactful than switching to a concrete or wood structural system. The benefit of an LCA study can be seen here, both in terms of quantifying (and disproving) assumptions as well as taking a holistic look at the impacts of the building's structure across several impact categories.

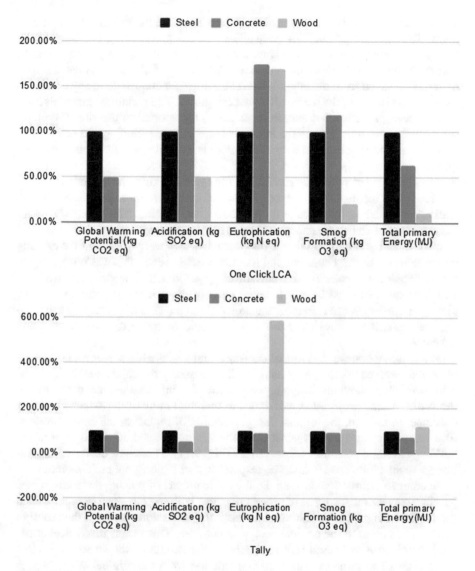

Fig. 9.10 Structural system comparison using Tally and One Click LCA comparing the original steel structural system, used as a baseline, to a concrete and wood option. Results are shown for life cycle stages A–C

9.2.3 Adaptive Reuse

While this project design does maintain the original columns from the existing building, the importance and potential of the existing materials were not fully realized during the design stage. Existing buildings are an invaluable resource when trying to mitigate environmental impacts. As the world continues to build at unprecedented rates, it is imperative for designers to use the resources available to them. Today, some of our most valuable resources are existing buildings (King 2017). In terms of life cycle thinking, adaptive reuse becomes a powerful tool to reduce the overall environmental impacts of a building. Not only does the salvaged material get saved from the landfill, but it also does not contribute to any cradle-to-gate impacts in any of the included impact categories. The reclaimed materials have already been manufactured and, in some cases, transported to the site. Therefore, their impacts are only seen in the later life cycle stages, including end of life. The power of adaptive reuse and salvaged materials as sustainability tools cannot be ignored.

After reconsidering Building 197 using life cycle thinking, the approach to adaptive reuse was reconsidered. The original warehouse, composed of tilt-up concrete wall panels, lacked expression in form and adequate glazing. The easiest way to fix these issues from a design standpoint was to remove the existing envelope and build a new one. In the initial design, the only elements that were used from the existing structure were the steel columns and foundations. However, this was not an appropriate decision to make from a sustainability approach. The original envelope, which makes up almost 14% of the total mass of the existing structure, was completely discarded. When reframing the project from the goal of preserving and reusing materials, it becomes clear that efforts should be made to save the roof and envelope.

While many believe that formal decisions and sustainable performance have to be at odds with each other, they can be at least partially decoupled and both can be prioritized throughout the design process. Using this information, and the results of the WBLCA, one project was redesigned to maintain more of the existing building envelope. In the end, the original design intention of replacing the entire envelope (0% saved) was compared against two others, which saved 50% and 75% of the building envelope assemblies, respectively (Fig. 9.11). These new design configurations save an additional 6% and 7%, respectively, of the original building mass.

In order to clearly demonstrate the life cycle effects of reusing materials, a new baseline was established that replaced the entire building envelope but maintained the original height and footprint of the building. Decisions made in the redesign were kept as close to the original design as possible. This means that a steel structural system was still used independently of the results of the structural system comparison. The same exterior wall construction from the original WBLCA was used anywhere that a new portion of the envelope was added. The thermal performance of the existing envelope was increased by adding a layer of metal studs, filled with mineral wool insulation, and gypsum board to the interior of the wall construction wherever the envelope was salvaged. Fenestrations were added to the salvaged envelope to meet the circulation and natural light requirements of the updated building program. Six of the loading doors were saved to service the building, just like in the initial design, and the remainder were replaced with fixed glass.

The study was carried out using Tally because there is direct support for salvaged material within the program. Currently, the only way to do this in One Click LCA is by using a "used product" profile and changing the transport distance to zero.

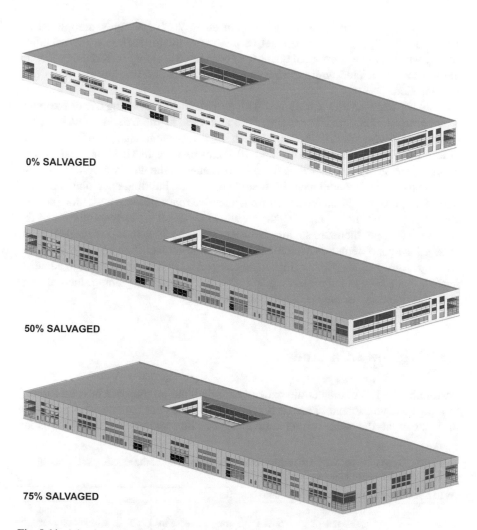

0% SALVAGED

50% SALVAGED

75% SALVAGED

Fig. 9.11 Adaptive reuse comparison comparing different amounts of the exterior envelope salvaged. The initial design intention, with 0% of the exterior envelope salvaged, was compared against an option with 50% and 75% of the envelope saved

Bionova is working on an easier way to incorporate existing materials in a study in future updates (Bionova 2020). When assigning a material in Tally, there is a box that indicates salvaged material. If this is checked, Tally does not include any impacts for the material at any life cycle stage. This simplifies the calculations, but is not indicative of real-world conditions. The impacts of transporting the materials to the construction site (if they are being salvaged from a different project), any necessary energy required to repair, refurbish, customize, or otherwise make the materials fit for purpose, as well as end-of-life impacts should be counted in the study. However, even with these impacts, it is possible to make significant savings compared to using new materials (Erduran et al. 2020). A study completed by Thormark (2000) shows approximately a 50% reduction in cradle-to-gate embodied energy when using a substantial quantity of reclaimed building materials, even with the impacts associated with refurbishing the materials and transporting them to a new location.

Looking at the results of the study with the baseline case having 0% of the envelope salvaged, compared to keeping 50% and 75% of the existing envelope, the impacts of saving materials can be seen in the percent change of the results (Table 9.2). Savings between 6% and 10% can be seen in all impact categories, except for ozone depletion where there is no change in the numbers. When looking at a comparison of a component LCA study of just the building envelope, inclusive of all fenestrations, the savings seen from adaptive reuse become a lot more noticeable (Table 9.3). When comparing the baseline new envelope to keeping 50% of the existing envelope, there are savings between 28% and 31% in all categories, again except for ozone depletion which has a negligible impact. Salvaging 75% of the envelope results in 36–40% savings in all categories. If the commitment to adaptive reuse is embedded in the project's goal and scope from the beginning, there is considerable potential for environmental impact savings.

9.3   Window LCA Study

Performing an LCA study at the scale of an individual product can be a useful way to mitigate environmental impacts that a WBLCA does not include. The advantage of a study at the scale of an individual building product is the widespread

Table 9.2 Comparison of WBLCA results with different amounts of the exterior walls salvaged in the context of the entire building mass

WBLCA study comparing salvage scenario benefits					
	0% Envelope salvaged (baseline)	50% Envelope salvaged	Percent decrease in impact	75% Envelope salvaged	Percent decrease in impact
Global warming (kg CO$_2$eq)	11,625,123	10,806,232	7%	10,585,924	9%
Acidification (kg SO$_2$eq)	61,480	56,419	8%	55,029	10%
Eutrophication (kg Neq)	2498	2339	6%	2289	8%
Smog formation (kg O$_3$eq)	695,131	642,559	8%	627,320	10%
Ozone depletion (kg CFC-11eq)	0.04	0.04	0%	0.04	0%
Non-renewable energy (MJ)	203,855,785	191,935,258	6%	188,666,546	7%

Study completed across life cycle modules A–D using Tally

Table 9.3 Comparison of only exterior wall LCA results (isolated from roof floor and structure) with different amounts of the envelope salvaged

LCA study on exterior wall components salvage scenario benefits					
	0% Envelope salvaged (baseline)	50% Envelope salvaged	Percent decrease in impact	75% Envelope salvaged	Percent decrease in impact
Global warming (kg CO$_2$eq)	2,641,678	1,823,964	31%	1,602,917	39%
Acidification (kg SO$_2$eq)	16,198	11,137	31%	9744	40%
Eutrophication (kg Neq)	577	417	28%	367	36%
Smog formation (kg O$_3$eq)	177,346	124,831	30%	109,540	38%
Ozone depletion (kg CFC-11eq)	0,009	0,007	22%	0,005	50%
Non-renewable energy (MJ)	37,851,425	25,979,769	31%	22,697,319	40%

Study completed across life cycle modules A–D using Tally

applications of the results. The results of a WBLCA study stop at just one building. However, any study at the scale of a product has the potential to have widespread impacts across many buildings. SolidWorks Sustainability is a very powerful tool to analyze an individual product or assembly in extreme detail. Due to the focused scale of a SolidWorks model, every single element of one building component can be modeled, as it would exist in the real world. Just like a BIM model, a SolidWorks model only contains as much information as a designer adds to it. A study at the scale of a discrete building component can be used to either compare existing, comparable products (i.e., wood window, aluminum window, clad window) or to optimize the design of a single product by changing the design (i.e., geometry, materials, finishes) while still meeting the same criteria (i.e., wind loads, water resistance, thermal performance) required of the product (Mirabella et al. 2014).

Taking the example of different window frame materials into SolidWorks Sustainability, it is possible to compare results across impact categories to find the best option. In this study, windows with a wood frame, aluminum frame, and wood frame clad in aluminum are compared. The FU of the study is one 2' × 4' fixed window with double pane glazing. The study assumes a 40-year service life and is inclusive of the material, manufacturing, transportation, and end-of-life cycle stages.

The wood window consistently had the lowest environmental impacts, followed closely by the aluminum clad wood window across the impact categories (Fig. 9.12). The results for water eutrophication are the closest for all three options, but these numbers are so small that the margins of error are higher and the results are less reliable. When looking at the differences between the three types of windows, it is important to remember that these are the impacts of just one 8sf window. However, the example of Building 197 has 14,765sf of windows across the project, and any savings get multiplied up to the scale of the building and beyond. There are

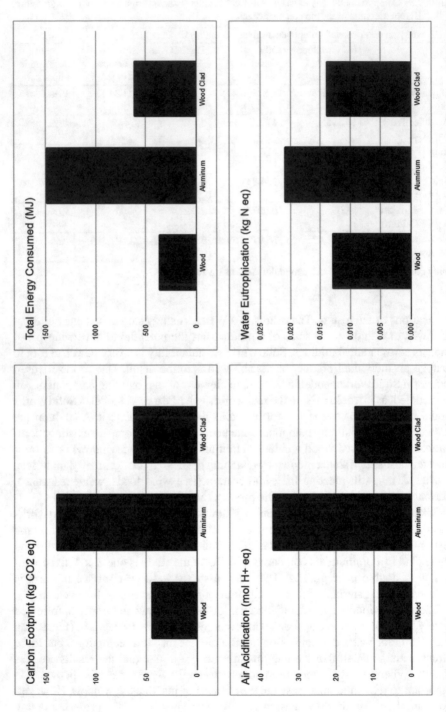

Fig. 9.12 Results of the window comparison study across the four life cycle stages included in SolidWorks Sustainability, inclusive of life cycle stages A–C

differences in performance and aesthetics of these three window options, and the environmental considerations cannot be the only deciding factor. However, the aluminum clad window could serve as a good compromise to having a similar look but better environmental performance.

9.4 Environmental Product Declarations

When working as an industrial designer or product designer steeped very closely in one specific area, the powers of LCA at the scale of an individual product are strong. However, someone who is working at a larger scale, such as an architect specifying products for a building, will not have the time or knowledge to fully conduct and utilize the results of such a study. In this case, an environmental product declaration (EPD) can provide insight to the designer. An EPD is a report generated that transparently reports the results of an LCA study for a specific product. They can help designers to evaluate comparable products and select the best one without having to complete the full LCA study themselves. However, inconsistencies in methodologies can make it so that the results of two EPDs are not comparable, and there are not publicly available EPDs for all products. Until EPDs are widely proliferated and standardized, studies like the example above is the most reliable way to compare individual products.

There are features specific to SolidWorks Sustainability that are not found in the WBLCA tools. Specifically, the ability to quickly assign and test different manufacturing locations and travel distances can be very valuable. SolidWorks Sustainability has information on different manufacturing conditions based on location, as well as default transport distances for products imported from varying regions. Tally and One Click LCA can also be used to test varying travel distances to simulate imported products, but it requires the user to enter information on the transportation distances for each product manually. Tally only has information on North American manufacturing processes, and while One Click LCA has databases for all over the world, they are covered under separate licenses, and so each project is restricted to data of one location. While location and transportation are important factors in a study, in most cases the impacts are dominated by the material stage.

9.5 Future Tools

At the moment, all of the tools discussed act independently of each other, instead of working in tandem. This lack of interoperability can severely limit the ways that designers can evaluate their projects based on a lack of access to software, knowledge of the software, or time. At the present moment, LCA, LCC, and operational energy studies must all be conducted separately, with few exceptions. SLCA tools and datasets are still on the horizon and active, and promising research by the LCA community continues.

Different tools evaluating the same criteria provide different relative advantages and disadvantages, each with its own internally consistent logic and frames of reference. In order for sustainability analysis to become fully incorporated into design workflows, interoperability and compatibility between programs are essential (Kneifel and O'Rear 2018). Initial steps toward this have been taken by some tools, but overall sustainability analysis requires the combination of many disparate programs and the careful monitoring for errors that may lie in the gaps between them.

In terms of current interoperability between evaluative criteria and tools, SolidWorks Sustainability and One Click LCA have started to bring LCC into the dashboard. Beyond that, BIRDS is the only tool that is bringing together all of the facets of sustainability, as defined by NIST, into one tool (Kneifel and O'Rear 2018). However, it lacks the ability to query user customized design options where other tools allow for analysis on a specific, unique design.

Another promising step toward interoperability is Rhino.Inside.Revit which helps to tie different CAD workflows into BIM-based LCA tools. The design consideration of ceiling height, presented in Sect. 9.2.3, could be fully parametric using a Grasshopper script, which would make the process of testing many ceiling heights quick and easy from a design standpoint. In using the WBLCA tools, each time a design parameter is changed, the user must go through and manually generate new results to the study, even though the material definitions did not change.

This makes the case for a fully integrated tool that provides live real-time feedback on design decisions. As more considerations—including LCA, LCC, operational energy performance, SLCA, and more—are combined into a single user interface that provides instant feedback with the design process, these tools become easier, even less time-consuming and less expensive for designers to integrate into their workflows than they are already.

References

Aktas CB, Bilec MM (2012) Impact of lifetime on US residential building LCA results. Int J Life Cycle Assess 17:337–349. https://doi.org/10.1007/s11367-011-0363-x

Athena (2019) About whole-building LCA and embodied carbon an Athena Institute briefing note. Athena Sustainable Materials Institute. http://www.athenasmi.org/wp-content/uploads/2019/09/About_WBLCA.pdf. Accessed May 2020

Athena (2020) Whole-building LCA guidelines and benchmarking initiative. Athena Sustainable Materials Institute. http://www.athenasmi.org/wp-content/uploads/2020/02/Athena_WBLCA_Guidelines_and_Benchmarking_Initiative_2020.pdf. Accessed May 2020

Bionova (2020) One Click LCA Spring 2020 online expert meeting

De Wolf C, Pomponi F, Moncaster A (2017) Measuring embodied carbon dioxide equivalent of buildings: a review and critique of current industry practice. Energy Build 140:68–80. https://doi.org/10.1016/j.enbuild.2017.01.075

Erduran DÜ, Elias-Ozkan ST, Ulybin A (2020) Assessing potential environmental impact and construction cost of reclaimed masonry walls. Int J Life Cycle Assess 25:1–16. https://doi-org. libdb.njit.edu:8443/10.1007/s11367-019-01662-2

Foundation Le Corbusier (n.d.) Maison Dom-Ino. http://www.fondationlecorbusier.fr/corbuweb/ morpheus.aspx?sysId=13&IrisObjectId=5972&sysLanguage=en-en&itemPos=102&itemCou nt=215&sysParentId=65&sysParentName=home

Herrmann IT, Moltesen A (2015) Does it matter which life cycle assessment (LCA) tool you choose? – a comparative assessment of SimaPro and GaBi. J Clean Prod 86:163–169. https:// doi.org/10.1016/j.jclepro.2014.08.004

King B (ed) (2017) The new carbon architecture: building to cool the climate. New Society Publisher, Gabriola Island

Kneifel J, O'Rear E (2018) Challenges and opportunities in quantifying and evaluating building sustainability. Technol Archit Des 2:160–169. https://doi.org/10.1080/2475144 8.2018.1497363

Mirabella N, Castellani V, Sala S (2014) LCA for assessing environmental benefit of eco-design strategies and forest wood short supply chain: a furniture case study. Int J Life Cycle Assess 19:1536–1550. https://doi-org.libdb.njit.edu:8443/10.1007/s11367-014-0757-7

Sinha R, Lennartsson M, Frostell B (2016) Environmental footprint assessment of building structures: a comparative study. Build Environ 104:162–171. https://doi-org.libdb.njit. edu:8443/10.1016/j.buildenv.2016.05.012

Speck R, Selke S, Auras R et al (2015) Life cycle assessment software: selection can impact results. J Ind Ecol 20:18–28. https://doi-org.libdb.njit.edu:8443/10.1111/jiec.12245

Thormark C (2000) Environmental analysis of a building with reused building materials. Int J Low Energy Sustain Build 1:1–18. http://hdl.handle.net/2043/9844

Appendix A

MIA project result comparisons over three WBLCA tools for multiple building construction assembly types, impact category, and life cycle stages as defined in standard EN 15978 – 8'-0" floor to ceiling.

© Springer Nature Switzerland AG 2021
J. Cays, *An Environmental Life Cycle Approach to Design*,
https://doi.org/10.1007/978-3-030-63802-3

	EN 15978 Life Cycle Stages					
Structure 8' fl.-clg.	**Athena**					
Impact Category	Product Stage A1-A3	Construction Stage A4-A5	Use Stage B2, B4, B6	End of Life Stage C1-C4	Module D D	Total A to C
	Concrete Structure					
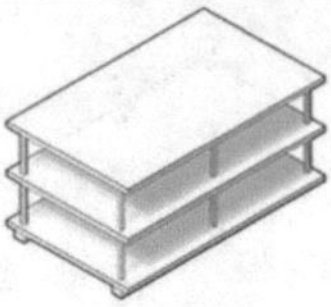 Global Warming (kg CO2)	28,800	2,670	0	1,710	2,970	33,200
Acidification (kg SO2 eq)	89	18	0	22	68	129
Eutrophication (kg SO2 eq)	27	2	0	1	0.4	31
Smog Formation (kg O3 eq)	1,500	531	0	704	69	2,730
Total Primary Energy (MJ)	249,000	29,700	0	25,300	13,600	304,000
	Steel Structure					
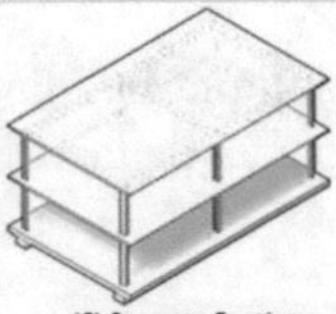 Global Warming (kg CO2)	22,900	2,020	0	1,280	-1,030	26,200
Acidification (kg SO2 eq)	80	18	0	15	-2	113
Eutrophication (kg SO2 eq)	17	2	0	1	-1	19
Smog Formation (kg O3 eq)	1,300	527	0	512	-0.2	2,340
Total Primary Energy (MJ)	245,000	24,700	0	18,900	-4,740	289,000
	Wood Structure					
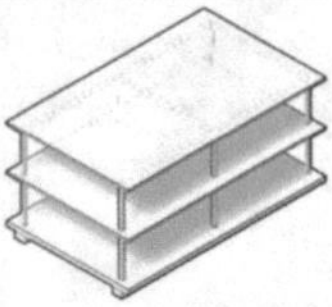 Global Warming (kg CO2)	12,200	2,810	0	1,220	-11,100	16,200
Acidification (kg SO2 eq)	53	27	0	16	0.2	96
Eutrophication (kg SO2 eq)	13	2	0	1	0	16
Smog Formation (kg O3 eq)	1,030	842	0	529	2	2,400
Total Primary Energy (MJ)	173,000	37,400	0	18,100	389	229,000

Concrete Structure labels:
(6) Concrete Footings
Concrete Slab on Grade
(6) Concrete Columns
(2) Concrete Slabs

Steel Structure labels:
(6) Concrete Footings
Concrete Slab on Grade
(6) Steel Columns
(14) Steel Beams
(2) Metal Deck Slabs w/ Concrete Topping

Wood Structure labels:
(6) Concrete Footings
Concrete Slab on Grade
(6) Glulam Columns
(14) Glulam Beams
(2) CLT Slabs w/ Concrete Topping

		EN 15978 Life Cycle Stages					
		Tally					
Structure 8' fl.-clg.	Impact Category	Product Stage A1-A3	Construction Stage A4-A5	Use Stage B2 - B6	End of Life Stage C2-C4	Module D D	Total A to C
	Concrete Structure						
	Global Warming (kg CO2)	56,090	453	0	4,065	-4,096	60,608
	Acidification (kg SO2 eq)	163	2	0	19	-7	184
	Eutrophication (kg SO2 eq)	10	0.2	0	1	-0.3	11
(6) Concrete Footings Concrete Slab on Grade (6) Concrete Columns (2) Concrete Slabs	Smog Formation (kg O3 eq)	2,987	69	0	374	-65	3,430
	Total Primary Energy (MJ)	520,873	6,582	0	69,592	-35,774	597,047
	Steel Structure						
	Global Warming (kg CO2)	38,092	405	0	2,200	-3,884	40,697
	Acidification (kg SO2 eq)	131	2	0	10	-7	143
	Eutrophication (kg SO2 eq)	7	0.2	0	0.5	-0.3	8
(6) Concrete Footings Concrete Slab on Grade (6) Steel Columns (14) Steel Beams (2) Metal Deck Slabs w/ Concrete Topping	Smog Formation (kg O3 eq)	1,963	62	0	202	-82	2,227
	Total Primary Energy (MJ)	390,621	5,890	0	37,661	-33,893	434,172
	Wood Structure						
	Global Warming (kg CO2)	2,131	542	0	17,859	-239	20,532
	Acidification (kg SO2 eq)	115	3	0	70	-12	188
	Eutrophication (kg SO2 eq)	7	0.2	0	16	-0.5	24
(6) Concrete Footings Concrete Slab on Grade (6) Glulam Columns (14) Glulam Beams (2) CLT Slabs w/ Concrete Topping	Smog Formation (kg O3 eq)	1,869	83	0	373	-132	2,325
	Total Primary Energy (MJ)	456,983	7,874	0	39,563	-76,557	504,420

		EN 15978 Life Cycle Stages					
		One Click LCA					
Structure 8' fl.-clg.	Impact Category	Product Stage A1-A3	Construction Stage A4	Use Stage B1-B6	End of Life Stage C1-C4	Module D D	Total A to C
	Concrete Structure						
	Global Warming (kg CO2)	25,142	5,283	0	2,370	N/A	32,795
	Acidification (kg SO2 eq)	64	7	0	9	N/A	80
	Eutrophication (kg SO2 eq)	23	4	0	2	N/A	29
(6) Concrete Footings Concrete Slab on Grade	Smog Formation (kg O3 eq)	1,020	65	0	226	N/A	1,311
(6) Concrete Columns (2) Concrete Slabs	Total Primary Energy (MJ)	128,352	80,467	0	58,165	N/A	266,984
	Steel Structure						
	Global Warming (kg CO2)	24,919	3,213	0	1,432	N/A	29,563
	Acidification (kg SO2 eq)	84	5	0	6	N/A	95
	Eutrophication (kg SO2 eq)	19	2	0	2	N/A	23
(6) Concrete Footings Concrete Slab on Grade (6) Steel Columns (14) Steel Beams	Smog Formation (kg O3 eq)	1,308	58	0	139	N/A	1,505
(2) Metal Deck Slabs w/ Concrete Topping	Total Primary Energy (MJ)	138,100	50,597	0	35,326	N/A	224,022
	Wood Structure						
	Global Warming (kg CO2)	16,481	2,571	0	1,135	N/A	20,187
	Acidification (kg SO2 eq)	79	5	0	7	N/A	91
	Eutrophication (kg SO2 eq)	17	2	0	10	N/A	28
(6) Concrete Footings Concrete Slab on Grade (6) Glulam Columns (14) Glulam Beams	Smog Formation (kg O3 eq)	517	72	0	202	N/A	792
(2) CLT Slabs w/ Concrete Topping	Total Primary Energy (MJ)	53,603	42,750	0	27,418	N/A	123,771

		EN 15978 Life Cycle Stages		
		Athena	Tally	One Click LCA
Structure 8' fl. - clg.	Impact Category	Total A to C	Total A to C	Total A to C
		Concrete Structure		
	Global Warming (kg CO2)	33,200	60,608	32,795
	Acidification (kg SO2 eq)	129	184	80
	Eutrophication (kg SO2 eq)	31	11	29
(6) Concrete Footings Concrete Slab on Grade (6) Concrete Columns (2) Concrete Slabs	Smog Formation (kg O3 eq)	2,730	3,430	1,311
	Total Primary Energy (MJ)	304,000	597,047	266,984
		Steel Structure		
	Global Warming (kg CO2)	26,200	40,697	29,563
	Acidification (kg SO2 eq)	113	143	95
	Eutrophication (kg SO2 eq)	19	8	23
(6) Concrete Footings Concrete Slab on Grade (6) Steel Columns (14) Steel Beams (2) Metal Deck Slabs w/ Concrete Topping	Smog Formation (kg O3 eq)	2,340	2,227	1,505
	Total Primary Energy (MJ)	289,000	434,172	224,022
		Wood Structure		
	Global Warming (kg CO2)	16,200	20,532	20,187
	Acidification (kg SO2 eq)	96	188	91
	Eutrophication (kg SO2 eq)	16	24	28
(6) Concrete Footings Concrete Slab on Grade (6) Glulam Columns (14) Glulam Beams (2) CLT Slabs w/ Concrete Topping	Smog Formation (kg O3 eq)	2,400	2,325	792
	Total Primary Energy (MJ)	229,000	504,420	123,771

Appendix B

MIA project result comparisons over three WBLCA tools for multiple building construction assembly types, impact category, and life cycle stages as defined in standard EN 15978 – 12′-0″ floor to ceiling.

© Springer Nature Switzerland AG 2021
J. Cays, *An Environmental Life Cycle Approach to Design*,
https://doi.org/10.1007/978-3-030-63802-3

	EN 15978 Life Cycle Stages						
	Athena						
Structure 12' Fl. - Clg.	Impact Category	Product Stage A1-A3	Construction Stage A4-A5	Use Stage B2, B4, B6	End of Life Stage C1-C4	Module D D	Total A to C
Concrete Structure							
	Global Warming (kg CO2)	31,800	2,890	0	1,840	3,910	36,500
	Acidification (kg SO2 eq)	101	20	0	23	9	144
	Eutrophication (kg SO2 eq)	29	2	0	1	0.5	33
(6) Concrete Footings Concrete Slab on Grade	Smog Formation (kg O3 eq)	1,650	590	0	751	91	2,990
(6) Concrete Columns (2) Concrete Slabs	Total Primary Energy (MJ)	291,000	32,300	0	27,200	17,900	350,000
Steel Structure							
	Global Warming (kg CO2)	23,300	2,050	0	1,300	-1,010	26,600
	Acidification (kg SO2 eq)	82	18	0	16	-2	115
	Eutrophication (kg SO2 eq)	17	2	0	1	-0.1	20
(6) Concrete Footings Concrete Slab on Grade (6) Steel Columns (14) Steel Beams	Smog Formation (kg O3 eq)	1,320	536	0	516	-23	2,370
(2) Metal Deck Slabs w/ Concrete Topping	Total Primary Energy (MJ)	252,000	25,100	0	19,100	-4,640	296,000
Wood Structure							
	Global Warming (kg CO2)	12,200	2,820	0	1,230	-11,200	16,300
	Acidification (kg SO2 eq)	53	27	0	16	0.2	97
	Eutrophication (kg SO2 eq)	13	2	0	1	0	16
(6) Concrete Footings Concrete Slab on Grade (6) Glulam Columns	Smog Formation (kg O3 eq)	1,040	848	0	531	2	2,410
(14) Glulam Beams (2) CLT Slabs w/ Concrete Topping	Total Primary Energy (MJ)	174,000	37,600	0	18,200	389	230,000

	EN 15978 Life Cycle Stages					
	Tally					
Structure 12' Fl. - Clg.	Impact Category	Product Stage A1-A3	Construction Stage A4-A5	Use Stage B2 - B6	End of Life Stage C2-C4	Module D D	Total A to C

Concrete Structure

	Impact Category	Product Stage A1-A3	Construction Stage A4-A5	Use Stage B2-B6	End of Life Stage C2-C4	Module D D	Total A to C
	Global Warming (kg CO2)	56,687	458	0	4,094	-4,184	61,239
	Acidification (kg SO2 eq)	165	2	0	19	-7	186
	Eutrophication (kg SO2 eq)	10	0.2	0	1	-0.3	11
(6) Concrete Footings Concrete Slab on Grade	Smog Formation (kg O3 eq)	3,018	70	0	376	-67	3,464
(6) Concrete Columns (2) Concrete Slabs	Total Primary Energy (MJ)	527,126	6,665	0	70,089	-36,541	603,880

Steel Structure

	Impact Category	Product Stage A1-A3	Construction Stage A4-A5	Use Stage B2-B6	End of Life Stage C2-C4	Module D D	Total A to C
	Global Warming (kg CO2)	38,788	422	0	2,200	-3,810	41,410
	Acidification (kg SO2 eq)	134	2	0	10	-7	146
	Eutrophication (kg SO2 eq)	7	0.2	0	0.5	-0.3	8
(6) Concrete Footings Concrete Slab on Grade (6) Steel Columns	Smog Formation (kg O3 eq)	1,981	65	0	202	-80	2,248
(14) Steel Beams (2) Metal Deck Slabs w/ Concrete Topping	Total Primary Energy (MJ)	399,686	6,137	0	37,672	-33,249	443,495

Wood Structure

	Impact Category	Product Stage A1-A3	Construction Stage A4-A5	Use Stage B2-B6	End of Life Stage C2-C4	Module D D	Total A to C
	Global Warming (kg CO2)	1,764	548	0	18,155	-215	20,467
	Acidification (kg SO2 eq)	116	3	0	71	-12	190
	Eutrophication (kg SO2 eq)	7	0.2	0	17	-0.5	24
(6) Concrete Footings Concrete Slab on Grade (6) Glulam Columns	Smog Formation (kg O3 eq)	1,882	84	0	377	-134	2,343
(14) Glulam Beams (2) CLT Slabs w/ Concrete Topping	Total Primary Energy (MJ)	461,627	7,969	0	39,725	-77,723	509,321

		EN 15978 Life Cycle Stages					
		One Click LCA					
Structure 12' Fl. - Clg.	Impact Category	Product Stage A1-A3	Construction Stage A4	Use Stage B1-B6	End of Life Stage C1-C4	Module D D	Total A to C
	Concrete Structure						
	Global Warming (kg CO2)	25,312	5,319	0	2,386	N/A	33,017
	Acidification (kg SO2 eq)	64	7	0	9	N/A	80
	Eutrophication (kg SO2 eq)	23	4	0	2	N/A	29
(6) Concrete Footings Concrete Slab on Grade (6) Concrete Columns (2) Concrete Slabs	Smog Formation (kg O3 eq)	1,027	65	0	228	N/A	1,320
	Total Primary Energy (MJ)	129,222	81,012	0	58,559	N/A	268,793
	Steel Structure						
	Global Warming (kg CO2)	25,992	3,228	0	1,438	N/A	30,658
	Acidification (kg SO2 eq)	89	5	0	6	N/A	100
	Eutrophication (kg SO2 eq)	20	2	0	1	N/A	23
(6) Concrete Footings Concrete Slab on Grade (6) Steel Columns (14) Steel Beams (2) Metal Deck Slabs w/ Concrete Topping	Smog Formation (kg O3 eq)	1,367	61	0	139	N/A	1,567
	Total Primary Energy (MJ)	138,101	51,022	0	35,485	N/A	224,608
	Wood Structure						
	Global Warming (kg CO2)	16,576	2,576	0	1,135	N/A	20,288
	Acidification (kg SO2 eq)	80	5	0	7	N/A	92
	Eutrophication (kg SO2 eq)	17	2	0	9	N/A	28
(6) Concrete Footings Concrete Slab on Grade (6) Glulam Columns (14) Glulam Beams (2) CLT Slabs w/ Concrete Topping	Smog Formation (kg O3 eq)	530	73	0	202	N/A	805
	Total Primary Energy (MJ)	53,603	42,897	0	27,535	N/A	124,035

Structure 12' Fl. - Clg.	Impact Category	EN 15978 Life Cycle Stages		
		Athena	**Tally**	**One Click LCA**
		Total A to C	Total A to C	Total A to C
	Concrete Structure			
(6) Concrete Footings Concrete Slab on Grade (6) Concrete Columns (2) Concrete Slabs	Global Warming (kg CO2)	36,500	61,239	33,017
	Acidification (kg SO2 eq)	144	186	80
	Eutrophication (kg SO2 eq)	33	11	29
	Smog Formation (kg O3 eq)	2,990	3,464	1,320
	Total Primary Energy (MJ)	350,000	603,880	268,793
	Steel Structure			
(6) Concrete Footings Concrete Slab on Grade (6) Steel Columns (14) Steel Beams (2) Metal Deck Slabs w/ Concrete Topping	Global Warming (kg CO2)	26,600	41,410	30,658
	Acidification (kg SO2 eq)	115	146	100
	Eutrophication (kg SO2 eq)	20	8	23
	Smog Formation (kg O3 eq)	2,370	2,248	1,567
	Total Primary Energy (MJ)	296,000	443,495	224,608
	Wood Structure			
(6) Concrete Footings Concrete Slab on Grade (6) Glulam Columns (14) Glulam Beams (2) CLT Slabs w/ Concrete Topping	Global Warming (kg CO2)	16,300	20,467	20,288
	Acidification (kg SO2 eq)	97	190	92
	Eutrophication (kg SO2 eq)	16	24	28
	Smog Formation (kg O3 eq)	2,410	2,343	805
	Total Primary Energy (MJ)	230,000	509,321	124,035

Appendix C

Percent increase in GWP and Acidification of the MIA exterior wall and structural types going from an 8′ to a 12′ floor to ceiling. Comparisons show general agreement between LCA tools.

© Springer Nature Switzerland AG 2021
J. Cays, *An Environmental Life Cycle Approach to Design*,
https://doi.org/10.1007/978-3-030-63802-3

Environmental impacts of various exterior wall types across three tools			
	Athena	Tally	One Click LCA
GWP - Percent Change in Impact from 8' to 12' Ceiling			
Masonry	N/A	35.96%	36.36%
CMU	35.22%	36.37%	36.36%
Precast Concrete	36.21%	36.36%	36.37%
Brick Cladding on Metal Studs	35.86%	35.54%	28.83%
Steel Panels on Metal Studs	36.01%	36.36%	36.35%
Brick Cladding on CMU	36.07%	36.31%	40.65%
Wood Cladding on Wood Studs	36.12%	36.95%	29.78%
Acidification - Percent Change in Impact from 8' to 12' Ceiling			
Masonry	N/A	34.85%	36.36%
CMU	35.48%	36.96%	36.41%
Precast Concrete	35.58%	36.17%	35.94%
Brick Cladding on Metal Studs	35.61%	34.48%	16.13%
Steel Panels on Metal Studs	35.87%	36.65%	36.66%
Brick Cladding on CMU	35.81%	35.37%	38.65%
Wood Cladding on Wood Studs	35.62%	36.11%	22.73%

Environmental impacts of various stuctural types across three tools			
	Athena	Tally	One Click LCA
GWP - Percent Change in Impact from 8' to 12' Ceiling			
Concrete	9.94%	1.04%	0.68%
Steel	1.53%	1.75%	3.70%
Wood	0.62%	-0.32%	0.50%
Acidification - Percent Change in Impact from 8' to 12' Ceiling			
Concrete	11.63%	1.09%	0.00%
Steel	1.77%	2.10%	5.26%
Wood	1.04%	1.06%	1.10%

Index

© Springer Nature Switzerland AG 2021
J. Cays, *An Environmental Life Cycle Approach to Design*,
https://doi.org/10.1007/978-3-030-63802-3

Printed in the United States
by Baker & Taylor Publisher Services